A LIFE UNDERWATER

Charlie Veron was born in Sydney and now lives near Townsville. He has three higher degrees in different fields of science and was the first full-time researcher on the Great Barrier Reef. He was the first scientist to be employed by the Australian Institute of Marine Science, becoming Chief Scientist of that organisation in 1997. He has over a hundred publications on all aspects of corals, from palaeontology, taxonomy and biogeography to physiology and molecular science, and has published widely on other subjects, notably evolution and mass extinctions. He has also established a major website at coralsoftheworld.org. His many awards include the Darwin Medal, the Australian Marine Sciences Association Jubilee Pin, and a Lifetime Achievement Award from the American Academy of Underwater Science. His work has underpinned most major reef conservation initiatives over the past two decades, inside the Coral Triangle and beyond. Charlie has been diving since the age of eighteen, logging six thousand hours underwater, and has participated in sixty-seven expeditions to most of the world's key reef regions.

ALSO BY CHARLIE VERON (AS J.E.N. VERON)

Scleractinia of Eastern Australia
(five volumes with multiple co-authors)

Corals of Australia and the Indo-Pacific

A Biogeographic Database of Hermatypic Corals

Hermatypic Corals of Japan
(monograph)

Hermatypic Corals of Japan
(book, with Moritaka Nishihira, in Japanese)

*Corals in Space and Time: The Biogeography
and Evolution of the Scleractinia*

Inge

Corals of the World
(book in three volumes)

Coral ID
(with Mary Stafford-Smith, CD-ROM)

*A Reef in Time: The Great Barrier Reef
from Beginning to End*

Corals of the World
(website with three co-authors)

CHARLIE VERON
A LIFE UNDERWATER

VIKING
an imprint of
PENGUIN BOOKS

VIKING

UK | USA | Canada | Ireland | Australia
India | New Zealand | South Africa | China

Penguin Books is part of the Penguin Random House group of companies whose addresses can be found at global.penguinrandomhouse.com.

First published by Penguin Random House Australia Pty Ltd, 2017

Text copyright © John Veron 2017

The moral right of the author has been asserted.

All rights reserved. Without limiting the rights under copyright reserved above, no part of this publication may be reproduced, stored in or introduced into a retrieval system, or transmitted, in any form or by any means (electronic, mechanical, photocopying, recording or otherwise), without the prior written permission of both the copyright owner and the above publisher of this book.

Cover design by Alex Ross © Penguin Random House Australia Pty Ltd
Cover photographs by Tristan Bayer, Charlie Veron and Stoneography/Getty Images
Typeset in Adobe Garamond Pro 12pt/16pt by
Samantha Jayaweera, Penguin Random House Australia Pty Ltd
Colour separation by Splitting Image Colour Studio, Clayton, Victoria
Printed and bound in Australia by Griffin Press, an accredited ISO AS/NZS 14001 Environmental Management Systems printer.

National Library of Australia
Cataloguing-in-Publication data:

Veron, J.E.N. (John Edward Norwood), author.
A life underwater / Charlie Veron.
9780143785460 (paperback)
Includes index.

Veron, J. E. N. (John Edward Norwood).
Marine scientists – Australia – Biography.
Marine biologists – Australia – Biography.
Coral reef conservation – Queensland.
Marine sciences – Queensland.

penguin.com.au

*For Mary, whose thoughtful
caring hand is in everything I write*

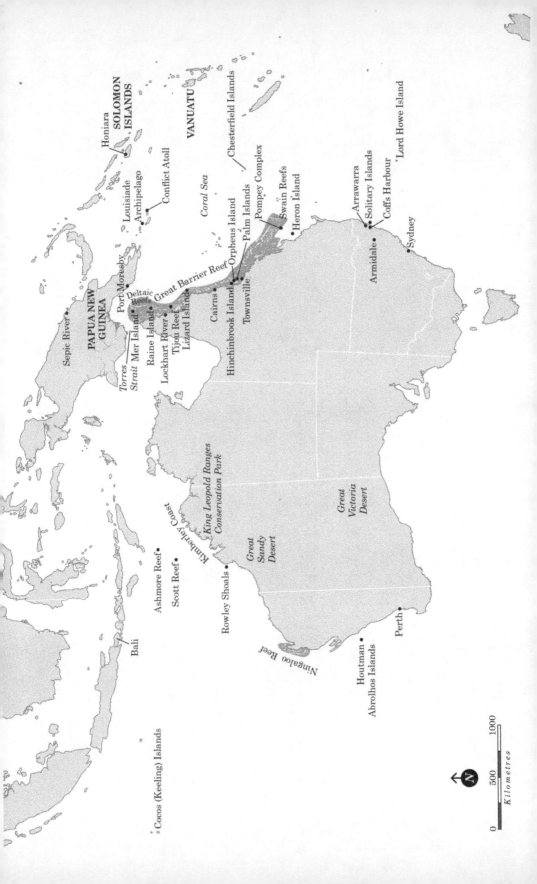

CONTENTS

Map of place names in Australia and the region vii
Map of place names in the rest of the world ix

Many Beginnings 1
A jar of worms *1*   Little Mr Darwin *9*   Him and me *21*
Freedom *38*   Delayed metamorphosis *49*   Scuba *57*
Serendipity *63*

The Great Barrier Reef 72
Matters of politics *72*   The tropics *76*   Underwater utopia *83*
Reef builders *88*   Expedition north *92*

A Wayward Career 104
The first AIMS scientist *104*   Russians for starters *109*
Big dusty museums *118*   Rivendell *129*
Monographs and bureaucracy *134*   The Reef expeditions *140*
Inge of Orpheus *149*   Two daughters *158*   Going west *168*

Travels Abroad 178
Trouble in Japan *178*   Pacific forays *189*   The Indian Ocean *197*
A sad end and a lifeline *205*

Time and Place 212
Invading deep time *212*   Hell's atoll *220*
A different evolution *223*   Origins of The Reef *233*
The Coral Triangle *240*

Big Pictures *245*
    A boy and a book *245*   Paths of conscience *249*   Reefs in time *253*
    A very big website *263*   Names that matter *266*

In Retrospect *277*
    The underworld *277*   Views from my coffin *287*

Afterword *294*

Notes 297
Acknowledgements 301
Index 304

Many Beginnings

A jar of worms

I can't say I was born a marine biologist but it gets close to that. When I was six my family rented a shack for the summer holidays at Collaroy, one of Sydney's northern beaches. A sandy-haired, blue-eyed ball of energy, I spent most days in the surf with my long-suffering mother, until one morning I caught a nasty dumper and had to be rescued. We abandoned the surf after that and strolled down the beach to Long Reef, which has a spectacular, wave-cut rock platform. At high tide this is a dangerous place. Waves echo around its outermost point where there's a slight rise, then come rolling in from both sides, leaping against each other in a panorama of angry white on churning blue. But that morning it was low tide and all was peaceful, so my mother left me to play in a small rock pool near the shore.

I saw plenty of anemones, some red, some green, with long tentacles surrounding what I supposed was a mouth. I held my finger close to a red one, then gave it a quick touch. The anemone didn't sting – I could stick my finger right into its belly – it just curled its tentacles in and then it looked like a little blob of jelly. That wasn't much fun, so I turned over a rock to find a worm and held it over another anemone. When I dropped the worm, the anemone grasped it in its tentacles. The worm thrashed about, struggling for its life. It almost escaped but the anemone held on, and then, with more tentacles to the fore, it started to pull the frantic worm towards its mouth. As the battle raged I watched the worm. It was just like a long thin centipede, with a body made up of segments, each with a little leg, but not jointed legs like centipedes had – these were legs that didn't seem to do anything. It had two long antennae on its head and a couple of shorter ones beneath, but more interestingly it had black eyes, as small as pin pricks, that looked as if they were painted on. This animal could see, but what? Then, as it twisted, I saw that it had jaws. They weren't small; surely the worm could chomp the anemone's tentacles off, but it wasn't trying to. Why not?

Earthworms didn't have legs or antennae or jaws, but I wasn't sure if they had eyes – I'd have to check that out.

I felt terrible when I realised that the anemone was stinging the worm to death, so I pulled it free, almost pulling it in half – but too late, the worm just lay on the bottom of the pool, only moving because ripples of water had started coming in. With a slightly guilty conscience, I dropped the poor dead worm onto another anemone, but surprisingly nothing happened. Did anemones only think worms were food if they struggled? I tried to work that out by twiddling tiny bits of kelp in the tentacles of another anemone, but again, no luck – they must be smarter than they looked. So I caught a little brittle star with long wavy arms and tried to get an anemone to attack that; still no luck. Did anemones think brittle stars weren't edible?

I looked about for something else to try and spied a green anemone growing on a cobble about the size of my fist. Very carefully I moved that up to a nearby green anemone and watched. Again nothing happened, they just ignored each other. Larger ripples were now invading the pool as the tide came in; I didn't have much time. I moved the green anemone up to a red anemone, expecting nothing to happen again, but war broke out almost immediately – the anemones started attacking each other. Through the ripples I could see that the green one was winning; the red one was withdrawing its tentacles, turning itself into a jelly blob. Were the green and red anemones not related? Was that why they fought? Would the red one be able to move away or would it be killed? My head was full of such questions, but my mother was calling. It was time for lunch.

The next day I couldn't wait to go back to Long Reef. The tide was out so off we went, my mum and I, armed with a bowl from the beach-house kitchen to put animals in. Finding a lot of them was easy back then, all I had to do was turn over rocks, the bigger the better. There were hundreds more brittle stars, each adorned with a different pattern, and everywhere were starfish in a multitude of colours, mini works of art. I turned one upside down; it waited a few minutes before extending tiny tube feet with suckers on the end and waved them about until one touched a rock. Then, with a lot more feet attached to the rock, it reached back a whole arm. With amazing strength it levered its body over. I turned another upside down in water that just covered it and watched it struggle until I felt sorry for it and gave it a helping hand.

There were hermit crabs of all sizes that vanished into their shells when I got too close. If I kept very still they would reappear, look around with tiny stalked eyes then move off, dragging their shell and leaving minute tracks in the sand. So many sorts of animals, but from what I could see the pools further out looked bigger and better.

The following day, my mother watched me trot down the sandy path to the beach, plastic bucket in hand. I also had a scoop net that my father had helped me make from a wire coathanger and a piece of mosquito net. 'Be careful, stay in the rock pools, and come back as soon as the tide starts to come in,' she called.

'And stay out of the water!'

There were many more interesting animals further out – several types of pretty sea slugs, more kinds of worms under rocks, including tiny tube worms with little doors on their tubes, and oysters all over the place. There were chitons with their eight plates of armour, limpets, barnacles, and purple urchins with spines always moving. I caught several different kinds of fish with my net, and crabs. Then I saw something moving slowly and got it into my net, and joy of joy, it was a blue and white sea slug, just about the most spectacular animal imaginable. To get a good look at it I took it to a little pool but to my dismay it managed to crawl under a rock. No matter, I'd keep a lookout for another.

I could see waves breaking on the outer point of the platform. They were a long way out but even in the distance looked really scary. Could I run all the way back to the beach from there when the tide turned? I sure could try. Out there I went.

Right at the water's edge there were beds of kelp, thick mats twisting and turning in lovely long curves as waves washed them back and forth. There was a mass of purple sponges under most rocks (were they animals?) and tiny tube worms. I was catching a blue-swimmer crab that was trying to bite me with its great big pincers when I heard someone yell, 'Hey boy, what ya doing there?'

It was a fisherman. He told me I shouldn't be out here because it was dangerous. I showed him all the animals in my bucket. He turned over a big rock for me and said, 'You gotta remember to turn 'em back, otherwise all those animals will die.'

I felt bad about all the rocks I'd turned over and just left.

The fisherman cut something brown off a rock. It looked like a dead tree stump but was smaller than my foot. He said it was a sea squirt, and made good bait. An animal stuck to a rock? What was he talking about? He threaded the guts of one onto his hook. It sure looked like bait.

Next day, the fisherman wasn't there. I had the whole place to myself. The waves swirled and churned back and forth, forming whirlpools with columns of bubbles disappearing into the depths, and every now and again a larger wave came right up, foaming and hissing, before gurgling back down to a ledge deep below the next wave. I couldn't see any bottom beyond that. I felt danger was everywhere, but I'd just spotted what I later learnt was a large mantis shrimp. I *had* to catch that, it was the most beautiful thing I'd ever seen. And I did, after a couple of plunges with my net. When I looked up it was right into the face of a wave looming above me. I dropped my bucket and net, grabbed some kelp and held on like hell. The wave tried to throw me up onto the rocks and then, worse, it tried to drag me off them. The kelp held on and so did I. As soon as I felt the worst of it was over I scrambled up to higher ground and looked back. My bucket was gone but my net was there, in the kelp right on the lip of the ledge. I scrambled down to get it before it was taken out by the next wave.

I met the fisherman once more that holiday. 'Hey boy, I thought I told you not to come out here. It's bloody dangerous. But before you go home, I found something for you.'

It was a jar. Not an ordinary jar, but one that had been pounded by sand and surf so long it was frosted all over. A treasure beyond description.

Back home my mother, thinking I'd been playing in a rock pool with my big sister Jan who she'd sent to keep an eye on me, agreed my jar was very pretty and told me it was a preserving jar. She bought a lid, rubber seal and metal catch for it. There were lots of similar jars in the hardware shop, but none all frosted over like mine. I kept it for years. It was my special jar for Long Reef treasures, especially worms, and that summer it was quickly filled. The holiday shack we rented didn't have many saucepans or bowls or jars and those we had were soon full of booty, some of which was starting to smell bad. My father bought me a bottle of methylated spirits that he said would preserve the animals, so I could sort them out and keep only the ones I wanted. This worked for a while, but the Long Reef rock platform is a big place and there was a lot of collecting to be done.

On my eighth birthday I was thrilled to be given an aquarium, and I discovered that living marine life is a lot more interesting than dead marine life. I kept small crabs, anemones, worms, brittle stars, chitons, sea cucumbers, sea urchins, starfish and sea shells, all easy pickings from Long Reef, as were barnacles – my favourites – once I found some on a piece of rock I could hammer off. I watched all these little critters in my aquarium, chin on hand from the back of a couch, for hours. My father found me a supply of brine shrimp eggs at an aquarium shop so I could feed my animals live food. I dropped bits of meat in and watched the crabs eat, their little mechanical mouths going flat out.

But despite my best efforts most of the animals didn't stay alive long: the water kept going bad and I couldn't work out why. The only other marine aquarium I'd seen was at Taronga Zoo, and they could pump as much seawater as they liked from Sydney Harbour. So I regularly asked my mother to drive me to nearby Roseville where I could get seawater from the small estuary in a jerry-can; I'd then have to find someone strong to help us lug it back to the car. As Roseville was no good for collecting, I also had to persuade Mum to take me back to Long Reef to stock up, over an hour's drive from home. I learnt that keeping a marine aquarium was a matter of trial and error, mostly error in my case, but one stock-up trip was a bonanza. I turned over a rock and there it was, a beautiful little octopus.

Heart pounding, I took my time to catch it, gently manoeuvring it with my hand until it darted straight into my net. What a prize! Ockie, as I called him, was instantly my most treasured possession. I cleared everything out of my aquarium to keep his water as fresh as possible. Feeding him was a matter of catching a ghost crab from the mudflats at Roseville. He had a voracious appetite, immediately pouncing on the poor crab and whisking it away to a little rock cave I arranged for him. Ockie also liked hermit crabs, gathering up all I gave him and leaving the empty shells in a neat pile by his cave entrance. Much as I tried, I couldn't see how he got the crabs out of their shells.

I soon found out that I didn't need to catch crabs at all; I could get a scrap of crabmeat or a prawn from our fish shop and feed him that. If he was hungry I would see him waiting for me at his cave entrance, usually turning pale tan as he searched around with his tiny hooded eyes. He quickly learned to climb up the glass and take his dinner straight from my fingers, and shortly after that I had him climbing out of the water and onto my arm. He would find his food and scramble back again on his eight little stick-on arms. He would always say thankyou by turning very dark and flashing spectacular blue rings at me before disappearing. I was convinced that Ockie learned to come when he was called, and he did so almost every afternoon for nearly a year.

I changed his water regularly and believed he would be my friend for life. Then, coming back from a family weekend away, we all noticed a familiar smell in the house. Ockie was already little more than a mound of decomposing sludge. I was devastated.

Curiously, there was nothing about the southern blue-ringed octopus in any of my books, so I didn't know then that these little creatures only live for a year or so. And nobody knew the danger I had been in. It wasn't until many years later that the *Sydney Morning Herald* ran a headline announcing that the southern blue-ringed octopus is one of the deadliest creatures on Earth, with venom from one bite enough to kill a dozen men. Since then its status has become legendary, helped along by the James Bond movie *Octopussy* and later by a Michael Crichton novel. Needless to say, Ockie never bit me.

His death had a continuing effect on me. It wasn't just that I'd lost a treasured pet, it started me thinking about all those rocks

I'd turned over at Long Reef and not turned back, about the hundreds of animals I'd collected and then cast off, and about the dozens more I'd tried to keep in my aquarium. Now they were all dead, and by my own hand. I decided that Ockie would be the last: if I was going to watch animals, I was going to watch them where they lived, and not kill them.

Some fifty years later I made a pilgrimage back to Long Reef, by then a marine reserve complete with guided tours. The place was nothing like I remembered it; there was hardly any animal life anywhere, presumably courtesy of hundreds of kids on school excursions. I came at low tide so I could walk out to the rocky point; it was about 300 metres from the shore, making a round trip of over half a kilometre to be negotiated in a narrow window of time before the waves started rolling in. The barnacles and tube worms were as I remembered them but there were only a few living oysters, and all that remained of the sea squirts, or cunjevoi as they're better known, were their gelatinous cases, cut open by fishermen for bait.

I watched the kelp writhing in mesmerising serpentine curves as the waves surged and ebbed and a shiver went down my spine at the memory of the little boy who once played alone in such an appallingly dangerous place.

Me around the age of six.

Little Mr Darwin

It was hard going back to school after such a holiday, but at least Lindfield East Public School was fun, especially with the teacher I had. Her name was Mrs Collins and being a real Nature lover she taught us all sorts of interesting things.

The best days were Tuesdays, when we had show-and-tell. Girls, the silliest of creatures, brought bonnets and dolls and other such junk. I brought really good stuff, mostly dead insects I'd collected, the best being some wasps pinned to a piece of cardboard. I gave a talk about them and then made the girls scream when I went to give them a closer look. With that sort of encouragement, and a slightly guilty conscience, I arrived one Tuesday with a live funnel-web spider in one of my mother's biscuit jars.

Mum definitely wouldn't like me doing this . . .

Apart from being deadly poisonous, as all kids know, funnel-webs are very nasty-looking creatures, black and hairy with big curved fangs. The Sydney funnel-web is the deadliest spider in Australia and possibly the world. It is also by far the most aggressive. And just as I'd hoped, my spider sent the whole class into turmoil. The girls all started screaming, some of the boys too. Mrs Collins was horrified. She confiscated the biscuit jar, spider and all.

'You are *not* to bring something like this to school! What would your mother say?'

Apart from such setbacks Mrs Collins was always interested in my collections, spiders and scorpions excepted, and sometimes she drew pictures of them on the blackboard with coloured chalk. Several other boys also brought interesting things to show-and-tell, but I usually knew more about them than they did. Not a great way to earn friends but it made an impression on Mrs Collins. She had nicknames for some of us; Mr Darwin was mine. I had no idea why.

One Tuesday I brought my jar of worms from Long Reef. As I opened the lid most of the girls scattered to the back of the classroom expecting something dreadful. They weren't disappointed: I had forgotten to keep the metho topped up and the classroom got a blast of rotten worm gas, enough to make a sailor throw up.

'Charles Darwin, get that out of here,' Mrs Collins yelled.

Lunch break soon followed and so did the retaliation of my classmates. Charles Darwin? Within minutes the other kids had converted that to Charlie. I wanted to fight the kid who started it but was soon surrounded by his many supporters. They taunted me. Char-lee. Char-lee.

Mustn't cry.

It was miserable being picked on like this, by all the kids, even some of the girls. It wasn't as if I was unable to defend myself, because I was a good wrestler. The school had a sandpit where I regularly battled an older kid who reckoned he took judo lessons. He usually won but the boys in my class had no chance against me. All the more reason, it seemed, to gang up on me now.

My father, on hearing this tale, gave an unsympathetic smile. 'You should be called Charlie,' he said, 'you're the great-great-great-great-grandson of Bonnie Prince Charlie.' I forget just how many greats it was.

'Really?' I had no idea who Bonnie Prince Charlie might be.

'Your grandmother was a Stuart. You're a direct descendant through her. Take a look at your toes; the second and third are webbed.'

I took off a shoe and sure enough my second and third toes were not fully separated. I hadn't noticed that before. Dad explained that this was the mark of the Stuart family, of royalty. Was I in line for the throne of England?

My classmates weren't as impressed by this as I was and kept calling me Charlie, after Mrs Collins, not the bonnie prince. I didn't like that, but most kids had names they didn't like and

I couldn't do much about it. Gradually, use of my nickname spread, including to adults, except for my parents unless in jest. John, my real name, has long been foreign to me, used only for my passport, credit cards and the like.

For my seventh Christmas, Santa, in whom I continued to believe as long as he did his thing, had given me a copy of William Dakin's book *Australian Seashores*. It was subtitled *A Guide for the Beach-lover, the Naturalist, the Shore Fisherman, and the Student*. That was me! Or rather who I wanted to be, and it was full of photos of Long Reef and all the creatures that lived there. A lot of the book was too complicated for me to understand, but the captions of the photos had names for most of the animals I collected. I soon had a name for almost everything and could impress everybody, especially myself, with my knowledge of what was what. I must have impressed Mrs Collins also. I was not in her class by this time, but we were buddies and so I took my book to school to show her.

Not long afterwards my mother said she wanted to take me to visit a woman at Sydney University who worked with Professor Dakin as his assistant. Mum dressed me in my best clothes, combed my hair, and off we went on the train. It took us some time to find the zoology department but when we did we were shown into a laboratory, there to await Isobel Bennett, whose name was on the title page of *Australian Seashores*.

'Charlie dear, how nice to see you,' she said, beaming at me as she came in with hand outstretched.

I was stunned. How did she know that the kids at school called me Charlie? I shook the woman's hand tentatively. She was nothing like what I expected. She wore an ugly white coat and dark-rimmed glasses and she stank.

'What have you got to show me?' she said.

'I have some worms and stuff.' I produced my jar.

'Hang on,' said Miss Bennett cheerfully, and off she went to get an enamel sorting tray, a magnifying glass and some tweezers. She tipped the contents of my jar into the tray – I'd made sure there was plenty of metho this time – and started sorting with her tweezers.

'You've got a lot of polychaetes here,' she observed.

'Yeah, they're my specialty really. I know most of them. From your book.'

She looked at me over the top of her glasses. 'You do?'

'Yeah, I had a *Terebella* too, but it fell to bits. But what I really want to know about are cunjevoi. You have different names for them in your book and I couldn't follow it.'

Miss Bennett put down her tweezers and explained that cunjevoi are ascidians, a branch of the tunicates, which are hemichordates, part of the evolutionary path that leads to the chordate phylum.

'What's a chordate?'

'Animals with backbones, like you.'

Now I was really struggling. 'Cunjevoi haven't got backbones.'

'No, but they have larvae, like tiny tadpoles, and these do have something like a backbone, although it isn't really. It's called a notochord.'

I can't believe this. This woman is weird.

My disbelief must have shown.

'I'll find you a slide and you can see for yourself with a microscope.' She disappeared again.

When I found out what a microscope was, I knew it was something I had to have. I could see anything with it. I was going to have to keep Santa going for at least one more year.

Our discoveries with the microscope went on until my mother, bored to tears and refusing to look another worm in the face, said it was time to go.

While I wasn't aware of it then, Isobel Bennett must have got quite a buzz from our meeting. *Australian Seashores* had only just been published and here was a small boy who was really into it. She gave me a bottle of formalin – the stuff she smelt of – and some collecting tubes. I never used either; I liked my jar and hated formalin. But all the same, our meeting was one of my life's defining moments: Isobel Bennett was to become one of the most influential people in it.

I was always talking to my mother about how much I loved God, for I knew that Nature was his work and that made him very special. Mum always agreed. Dad never seemed to have anything to say on the subject and, poor man, always had something urgent to do on Sunday mornings. So it was usually just Mum, my sister Jan and I who went to church – St Alban's Church of England, near Lindfield train station. I would polish my shoes until they glowed (God noticed these things), plaster my hair down with water, and hold my chin high while Mum tied my tie.

I liked the hymns best. Mum sometimes whispered, 'Sing up,' which was hard to do if I had to read the words at the same time, but if I knew them by heart I would snap the hymn book shut and yell, 'Onward Christian soldiers, marching as to war,' so loudly and so high that people would turn and look.

God would approve too.

Not so Jan, who cringed with embarrassment and told me to shut up, digging me in the ribs. When that didn't work she took to sitting in the pew behind me so she could bash me over the head with her hymn book. I usually spent a couple of hours back at church in the afternoons at Sunday school. I was sure I was the deacon's pet pupil, the way I could recite more of the Bible than any of the other kids, and I knew all the saints and their whacky stories, and of course I knew everything about Jesus and the things he'd done.

In God's name, I saved all the pennies I could spare from my sixpence-a-week pocket money, putting them in a special tin moneybox. When it was full I proudly presented it to the deacon, for the wonderful job his church was doing rescuing poor black babies from their black heathen mothers and giving them to good white families for a proper Christian upbringing. These babies are now known as the stolen generations and I doubt they would thank me for helping finance their plight.

My God was the creator of the wonderful world of Nature, but unfortunately the deacon never seemed to talk about that. It was hard for me to see the connection between the church, with its fancy furniture, and the bush. My mother appeared uninterested when I pointed out that Noah's Ark was the most important story in the Bible. No matter, she knew best and she was the person I wanted to please.

I did not have the emotional contact with my father that I had with my mother. Mine was a postwar world where children were mostly women's business. Nevertheless Dad did a lot of things for me, including making a large pond for goldfish when I was about five. The prize of my pond was a rotund, goggly-eyed goldfish called Chloe. She was, I suppose, a good-looker as goldfish go, having lots of spots and a long graceful tail. She won a prize at a pet show, I can't remember for what, and went on living in that same pond for twenty-three years.

My father taught me how to make things and use tools, turning me into a handyman with do-it-yourself skills that I've used all my life. Aged eight, I made a pond myself, for freshwater turtles, out of an old copper clothes boiler. There was an area for the turtles to come out to feed on pieces of meat or bake in the sun. Unfortunately they regularly escaped and would be gone for days, until the whole family mounted a search party and caught them again.

After that, ponds became a way of life. The goldfish would breed profusely come early spring, and I fed them on powdered whale meat that Dad bought somewhere. Whale meat sold as pet food? I now wonder about that.

My father also made me a large aviary for finches, canaries, and sometimes budgies, which I bred. I made a run for rabbits and another for guinea pigs. He eventually got cranky about so many ponds and cages cluttering up his well-kept garden and wrecking his dahlia bed, but by then it was too late to do anything about it. Mum insisted I keep the boy guinea pigs separate from the girls, and of course I wondered why.

'Because we don't need any more guinea pig babies,' she said.

'Do guinea pigs need to get married to have babies?'

'Sort of.'

'How do humans get babies, Mum?'

'They're given by God, darling, when their mothers are married. Now run off while I make dinner.'

I found this unbelievable, at least as far as guinea pigs were concerned. I was nothing if not observant and knew that guinea pigs had babies after eating nasturtium leaves and that they enjoyed nasturtiums so much they always played games, like chasings and piggy-back, after a feast. This theory embarrassed my mother, who told me not to be silly. So with that encouragement I collected a large pile of nasturtiums, as I had many times before, from our neighbour's backyard where they grew like weeds. I piled them next to the guinea pig runs, then let them all out for a feast. Sure enough, a couple of months later, baby guinea pigs were everywhere. My first scientific experiment had proved my point.

I discussed this with Rick Smyth, who lived nearby and was in the class above me at school. Rick was the closest friend of my childhood. He was much more sensible than me, a counter to my constant excesses.

'Nasturtium leaves? That's dumb. Only guinea pigs like them – what about all the other animals? What about humans?' Rick had been doing some research with the aid of his dad's *Lilliput* magazines.

Curious, I went back to my mother, who declared I shouldn't talk about such things until I was grown up. And that was all the sex education I ever got, Rick too for that matter. So we kept up our research and came to the unavoidable conclusion, and like most kids of my generation, I couldn't believe it. My dad did *that* to my mum?

When I was almost ten my mother's black cocker spaniel, Lucifer, died. For me his death was almost as bad as Ockie's but at least it gave me a chance to campaign for a dog of my own, and after much opposition and negotiation it was agreed: for my tenth birthday I could go to the council pound and choose a dog, as long as it was small. I chose a very young puppy, one that was indeed small except for his feet, which were decidedly oversized. I named him Jinka and much to my delight, and my poor parents' dismay, he grew and he grew and he grew, until he took on the appearance of a cross between a red setter and a labrador. By the time he was a year old we all saw that Giant was his middle name and great dane was part of his pedigree.

From my childhood to university, Jinka was my constant companion and I loved him more than anything else in my life. I taught him to wrestle, getting him in a headlock and then onto his back, forcing his front paws to the ground. I did that until he reached full size and became strong enough to toss me off unless I put all my weight right on his paws. Our fights usually started with me giving his nose a slap. He would growl, bare his teeth, raise his hackles and launch himself at me. Other times he'd walk up to me and start growling: Want a fight? If I took off his collar he knew that meant I couldn't use it to cheat and so would have an especially hard time getting him onto his back. Then he would put up a hell of a battle; we could have been a stunt pair. My poor mother was always mending torn shirts and trousers and sticking bandaids onto cuts from accidental encounters with big teeth.

Perhaps it was because Jinka was the best bodyguard any child could have that my mother gave me a free rein to come and go just about anywhere I chose. I was always safe with Jinka and I chose to spend a lot of time in the bush, exploring the paths of Ku-ring-gai Chase, a rugged, beautiful national park about the size of Sydney that extended almost to our house. Any time I could spare I rode my bike to the bush and headed off down one of the tracks, Jinka racing beside me. Then we'd leave the path and walk on, away from all signs of habitation until we were completely alone.

Day or night, sun or rain, I loved being alone in the bush, and still do. Armed only with a knapsack, a couple of sandwiches and maybe an apple, I would go with Jinka for miles, until we reached a stream or some such peaceful spot, there to stop and sit, or just muck about. At these times I felt so much at peace that I would stop thinking. I wouldn't go to sleep, for I could still hear and see things around me, just not think about them. I might see a dragonfly hovering, then darting off to chase another away. I would feel the trees move in the breeze, perhaps hear a bird fly past. Ten or twenty minutes might go by. Then I'd vaguely wonder about something I saw or heard and, as if coming out of a trance, I would start to think again and return to my usual, worldly self. These episodes, which I assumed were normal, always left me relaxed yet invigorated, and with the feeling that I had been close to Nature, close to my God.

I now know this state is called meditation by some, a state of mind developed with practice, but for me it was completely effortless, something I slipped into naturally, and I often did so well into married life. I valued it as a gift. Later, there never seemed to be the time for it and eventually I forgot about it. Now I yearn for that peace and solitude, the tranquillity I once had.

One afternoon in August 2015 I was sitting on a beach being interviewed with an Aboriginal man, Jarmbi Githabul, for a video about climate change and what it meant for humanity. On my mention of meditation he smiled at me and told me to google 'dadirri'. 'You can call it deep listening,' he said. Later, I was astonished to read about dadirri, for that is exactly what I did and felt. Aboriginal people *feel* where they are, feel their land around them. I'm like them; I was born like them. Maybe all kids are, they just lose it before they know they have it. The freedom I had to roam at will in the bush during my childhood remains something precious beyond measure to me.

We stopped going to Collaroy around the same year I got Jinka, and went to other places for family holidays, including the Blue Mountains, a few hours west of Sydney by steam train. Our first trip was in midwinter, so cold we could have snow fights, but the attraction for me was a 'museum' next to our hotel. Actually it was a long fibro shed, owned by a local celebrity called Melbourne Ward, and unlike most modern museums it was designed to impart wonder rather than teach. It was crammed with natural history collectables of all sorts, with simple labels like 'jaw of great white shark', 'Triassic trilobite', or 'shrunken head, Papua New Guinea', but it was the fossils that interested me most. What had the world been like when they were alive, I wanted to know. And how long ago was that? Mr Ward always seemed to have the answers to my questions. I got to know him and he never wanted me to pay for my visits.

'You're a strange boy,' he once said, presumably referring to my love of his museum. Strange or not, I always came away with a thirst for knowledge.

My passion for Ward's museum must have struck a chord with Dad, for after my last visit he did a truly wonderful thing for me: he built a garage-sized 'museum' of my very own in the corner of our back garden, out of wooden packing cases that had been used to import cars from England. I loved my museum and kept all manner of exotic creatures in it: collections of dead insects and spiders as well as a live carpet snake, some lizards, axolotls and a big green frog. I also had a collection of fossils, minerals, rocks, and generally interesting things which I would add to whenever I could, borrowing them on a permanent basis from the attics or other repositories of junk that family friends had.

It wasn't long before I found another funnel-web spider, a very big female I called Spooks, short for Spooky Spider. I had a leaky old aquarium in the corner of my museum that I half filled with moist rotted vegetation from Dad's compost heap, and into it Spooks went. To my enormous delight she soon made herself a silken home, a beautiful funnel with more silk carpeted around her entrance; this had trip lines in it. To feed her I just dropped a cockroach into the aquarium and watched. As soon as the hapless insect trod on a trip line Spooks was on it in a flash, enveloping it in her big hairy legs and biting it with her long gruesome fangs. She would tackle anything, even another funnel-web, which were easy to find if you knew where to look. The only prey she seemed frightened of were bull ants. I was careful with these – they were big, very aggressive, and could sting like hell with both their abdomen and jaws at the same time. Maybe some primordial instinct warned Spooks that where there was one bull ant there would be many: she would rear up, fangs open, and back away. The ants too would rear up, jaws open, ready for battle.

One day I came across a large scorpion in a rubbish pile and manoeuvred it into a jam jar. I held the jar over Spooks's aquarium but couldn't make myself tip the scorpion in. What if it managed to sting Spooks? I couldn't bear the thought that she might get hurt. That night my mother woke me and held me close until I calmed down. I'd been dreaming about the battle and somehow I was in it, with Spooks, bigger than me, attacking me. Mum was horrified (by then a rather constant state of affairs), and when I got up next morning my father had found Spooks another home.

Still, I had to admit I had an enviable set-up, with Jinka's kennel next to my museum, rabbit and guinea pig runs, an aviary, fish ponds galore, and two aquaria that I used for breeding freshwater fish. On top of that, my father had kept *National Geographic* magazines ever since I could remember, even cataloguing the articles. With these at hand, and books from wherever I could find them, I began reading anything and everything I found interesting. I became a bookworm, and I still am: this was the beginning of another lifelong passion.

While there were no serious downsides to this life, I was aware that I kept having episodes of being alone inside my head, and these were becoming increasingly frequent. This was very frustrating for other people and sometimes it wasn't much fun for me either. I was repeatedly missing out on things I wanted to do because I didn't hear my name being called, and most days I was reprimanded at school for not paying attention. But I was doubtless more dreamy at home than anywhere else. 'Knock knock, anyone there?' Jan would say, often several times before I heard. I'd be far away in a world of my own, perhaps wondering why trees are green or how hot the sun is.

My poor sister, she was always on her guard with me. She had no liking for the insects and spiders I collected and was positively traumatised by live cicadas, grasshoppers and praying mantises. I would terrorise her when needs be, or for no reason at all, chasing her around the house or up the street, creature in hand. Even when I wasn't thus armed she was unable to defend herself against me; I would always manage to beat her in a wrestle, even when I was half her size. Yet she unfailingly came to my defence when I was in trouble, a very loyal sister.

When I was still in my tenth year my father suggested I might like to go to an abandoned quarry at Brookvale, near Sydney's northern beaches, to look for fossils. What a day that was. We found plants – including *Glossopteris*, a rainforest tree that once covered Gondwana before it broke apart into today's familiar continents – and parts of insects, and, best of all, a perfectly preserved fish. This opened a window on a world I knew too little about. I couldn't wait to show these spoils to my uncle Harold, curator of palaeontology at Sydney's Australian Museum, and the only one of my family whose interests were close to mine. As a young man, Harold Fletcher had journeyed with Afghan camel drivers in central Australia, and he'd also been – how I envied him – a member of Mawson's 1929–31 expedition to Antarctica (about which he later wrote a book, *Antarctic Days with Mawson*). I started regularly meeting up with Uncle Harold in the museum's storage rooms, where fossils were kept and where I could explore at will. He occasionally gave me specimens that were of no use to him but which were treasures for my own museum. And so from an early age, the main geological intervals of our planet were as familiar to me as local shops were to most other people.

Him and me

Barker College is one of Sydney's well-to-do private schools, a boys-only school when I was there. Mum's brother, my uncle John, had been head prefect there and later became chairman of the school's council. Although she never said so, my mother would have felt obliged to send me to Barker. Uncle John's son would be going there, and my name had been on the school's waiting list since I was born. I was enrolled in the junior school in 1955, at the age of ten, and was immediately advanced a year, to the second-last year of the junior school. I have no idea why, except that I was bigger than most kids my age. I was the youngest boy in my class but I was still one of the biggest and probably strongest – I was an early developer physically, but not in any other way.

Barker was nothing like the school I had just left. For a start, we all wore uniforms and were expected to do exactly as we were told. I got off on the wrong foot with my teacher from the beginning and repeatedly spent lunch breaks in detention for being late, for being untidy, and most particularly for not paying attention in class. There were only a few new boys in my class that year and we tended to gravitate together for company. Not that there was anything wrong with the other kids, it was just that they had their own social groups and we weren't in them. That didn't matter much – I wasn't a very social child anyway – but what did matter was a group of big kids in the class above who were always bullying others, especially the smaller boys in my class.

One lunch break I saw these big kids huddled on the ground, bellowing with laughter. I was curious and went over to see what was so funny. One had a magnifying glass, and by focusing the sunlight was using it to burn ants.

'That's cruel,' I shouted.

'Bugger off, Veron – prick.'

The following day at morning break they decided to have some more fun, this time with me.

'Hey Veron, come and look what we've got,' called the magnifying glass kid.

He had a cicada and was pulling its legs off one by one, as the little insect croaked in agony and beat its wings frantically. I was dumbfounded at such cruelty, then blind furious. I grabbed this horrible boy by the hair with one hand and let him have it with all the force I could muster with the other, right on the nose. Screaming in pain, he threw himself at me, and his mates immediately started chanting, 'Fight, fight, fight.'

I had never been in a fist fight before, and no doubt things would have gone badly for me had not my assailant tried to grapple with me. Big mistake. Wrestling was my thing and he quickly found himself with his face in the dirt and me on top of him, trying my absolute best to do to his arm what he had done to the cicada's legs. His screams for help were answered, not by his mates, but by the headmaster. 'Veron started it!' went the chorus, and I was taken by the ear to the school office, to receive a caning the likes of which I had never imagined. Worse, I was ordered straight back to my classroom, blubbering with pain, misery and the injustice of it all. There I was prodded with rulers every time the teacher turned to the blackboard, and I copped whispers of 'crybaby' from the boys nearest me.

My persecution stopped when the lunchtime bell went and the whole class found out who I'd been fighting with, and that I had very definitely won. Even so, the reputation I gained wasn't good – punching on the nose is about as taboo as playground rules get. I became a kid more to be avoided than befriended.

Bullying at Barker was so prevalent it was a way of life, yet the school turned a blind eye to it, all in the name of 'turning boys into men', so my father said. I am an emotional type and quick to anger. Bullying for me is cruelty, and cruelty, be it to animals or children, is the one thing that can make me completely lose my temper. Inevitably this soon happened again. A boy in my class was making a sickly little kid grovel for fun; my intervention turned into a very nasty business and this time I felt sure I'd be expelled. But no teacher arrived and, incredibly, nothing more came of it.

Apart from bullying, my main gripe with the school was that it radiated a creed of 'do what you are told, think what you are told, learn what you are told'. That, for me, was not just difficult, it was impossible, although at the time I didn't know it. I did try. I tried for years. And so the boy who tried gradually morphed away from the boy who was, until the divergence became so great they were two different kids altogether; they just occupied the same body. I came to think of the boy who tried as 'him', and the boy who was as 'me' – my real self. What I didn't realise then was that 'he' was a certain loser. My self-esteem whittled away and I retreated even more into my inner self, where at least there was something I could hold onto, perhaps even like.

By the time I was thirteen things were pretty bad. I had developed asthma, then considered a social disease of the weak, and it steadily worsened. The afternoons were the hardest; even walking slowly home from the train station was sometimes an ordeal, especially if I had books to carry, and often I had to stop and sit by the road, gasping for air. My doctor gave me some adrenaline pills for emergencies – those episodes when I felt seriously frightened – but all these did was make my heart race. Nevertheless, at school I was ordered onto the playing fields as was every other boy: for football, cricket, swimming, even athletics. The resulting humiliation was relentless. I couldn't do anything that involved running – puffers had not yet been invented – and I never learnt how to catch a ball.

To top it all off I developed a stutter, another social disease. By the time I was midway through school this was so bad I could hardly say anything. Sometimes a person I was trying to talk to found my struggle – eyes down, strain written all over my face – embarrassing and filled in the end of a sentence for me, or tried to. I got into the habit of rehearsing the next sentence in my mind, looking for trip words with a nasty initial consonant, and often I changed the word or the whole sentence, even if it meant saying something I hadn't intended. Anything to get the ordeal over with.

Often as not my name became Cha-cha-Charlie, yet curiously enough the bullies of the junior school kept their distance; any one of them could easily have taken his revenge. If that was a plus, it's the only one I can remember.

Most teachers ignored me or just handed out detentions for daydreaming, not paying attention or not doing as I was told. I was frequently caned for the same reasons. In all my eight years at Barker, only the biology teacher, Jim Bradshaw, seemed to think I knew more than I appeared to about whatever it was he was teaching. Occasionally, when he wasn't sure about something, he'd glance at me in class.

'The phylum Mollusca includes a wide range of animals such as gastropods, cephalopods, nudibranchs, chitons, limpets, snails and,' with a hesitant glance my way, 'barnacles.' If it happened that he was wrong, and it rarely did, I would shake my head slightly. 'Er, not barnacles, which are of course arthropods that look like molluscs.' Minuscule incidents, but very important for me at the time – something to cling to. I didn't always score a point. To most people I was a sort of retarded know-all, and nobody likes know-alls, especially teachers.

'Marsupials, as you know, are only found in Australia and, you may not know, New Guinea,' said Mr Bradshaw.

That was too much. I raised my hand as high as it would go.

'What, Veron?'

'That's n-not t-true, sir. They also oc-oc-occur in, in Am-America, sir.'

'Not true, eh? Turn to page sixty-seven and read what it says.'

'The b-book's wrong, sir.'

'Really, Veron? How do you suppose marsupials got to America? Except of course for the flying kangaroo,' he said, referring to the well-known icon of Qantas. That brought chuckles all round.

'They m-must have b-been there ever s-since Australia and Sou-Sou-South America were ja-joined to Ant-Ant-arctica.' That brought loud laughter.

'Continents floating about? Very funny, Veron. Take a Friday detention for wasting the class's time.'

I don't know why I knew about continental drift, except that such knowledge would have been normal for me. I remembered just about everything I found in *National Geographic* magazines or any library book I found interesting. I also loved reading my father's encyclopaedia, which I did for hours on end, something that worried my mother, probably because it had drawings of male and female genitalia in it.

Whenever I could I took revenge on my school the only way possible, by subterfuge. After 'Loco' Lathum, my near-senile chemistry teacher, had finished giving me a particularly harsh dressing-down, I snuck back to his lab and joined a bench gas tap to a sink water tap with a Bunsen burner tube and left them both running for the weekend. That put Loco's lab out of action for a week.

'M-me, sir? I d-d-don't know anything about ta-ta-taps, sir.' That he could believe.

A few months later someone glued the bastard's desk drawer shut . . .

The saddest part about all this was that my mother clearly believed what the school said of me, and my annual report cards were relentlessly condemning. I usually failed most subjects, I didn't pay attention in class, my homework was poor, and I had a bad attitude. Then my mother would remind me of the sacrifice my father was making in sending me to such a good school. I sometimes got self-defensive about all this; it wasn't fair.

I know a lot about most things. And I do try.

I couldn't explain what was wrong because I didn't know what was wrong. All I could offer, to my parents and myself, was that I couldn't help switching off more than most kids. I would stop listening to whatever the teacher was on about and wonder about something different, something interesting. Sometimes it was sparked by the teacher, but more often it was something internal, a return to a previous meandering. I was a square peg in a round hole and that was something the school had no means of understanding, nor any desire to. No doubt I wasn't the only boy in this sort of predicament but it usually seemed so.

Punishment rained down from every direction. Of all the subjects I hated, maths was the worst. One day my maths teacher, 'Cupie Booth', a creature so ancient he was old when my uncle went to Barker, made me write out Pythagoras's theorem fifty times for getting it wrong. The following day he told me to recite it in front of the class.

'P-P-Pythagoras's th-th-theorem,' I said in a haze of bewilderment, 'th-th-the sum of the s-square of the h-h-hypotenuse – is th-the square of of of the other sides.'

'Veron – outside.'

We all knew what that meant. If Bert Finlay, the deputy headmaster, a grim-faced portly ogre who patrolled the corridors dressed in a dark suit and waistcoat, saw a boy standing outside his classroom he was caned; no questions asked, no discussion. That was my fate many times and the canings got harder as the years dragged on, my backside becoming repeatedly decorated with black and blue stripes.

My father noticed them once, while I was undressing for a shower. 'Where did you get those nice little stripes from?'

'I, er, f-fell over, Dad.'

I suppose he felt that something had to change the way I was, and that caning might do it, although he never hit me himself. But nor did he ever take my side when I complained about my treatment at school.

When my father was sixteen he had lied about his age in order to enter the Royal Military College, Duntroon, a tough place then, where showers froze in winter and where, so he said, he could crack a walnut with his biceps. It seemed to me that he was a born soldier. He graduated a captain, with an engineering degree and all sorts of accolades. It wasn't long before he saw action in Afghanistan with the Australian army. I don't know what my mother saw in him, apart from him being good-looking and good at just about everything, but on his return to Sydney in

1934 they were married; she had only just left school.

Dad had a heroic career during World War II and became the youngest brigadier in the British Commonwealth, receiving two OBEs among a string of other medals. He saw action in North Africa and Asia, including, as a colonel, the Siege of Tobruk in 1941, when Australian and British forces repelled German panzer tanks in the first major defeat of the Germans, and then in Singapore and Papua New Guinea, fighting the Japanese. He sometimes talked to me about his exploits and became emotional about his wartime comrades. All this made a deep and lasting impression on me.

Poor Dad, during my school years he must have been wondering what he'd sired: a sickly boy who couldn't play any sport, who was a social misfit, who did badly at almost everything, who couldn't even talk properly. Not exactly a son for a war hero to be proud of.

One Friday afternoon in February 1959, my declining status at school took a dramatic turn for the worse. One of my classmates gave me an article torn from a glossy magazine. It was called 'A Missing Link?' and was about Darwin's theory of evolution. I read it, reread it, and then just sat gazing out my bedroom window for a very long time. The article featured a picture of a gorilla with eyes so deep they seemed to hold all the wisdom of the Earth.

Was this why Mrs Collins had nicknamed me Mr Darwin? Who was he anyway, and what did he have to do with me? Was I so special that God would give me everlasting life and the gorilla nothing? Did I have a soul – as I was always being told – and the gorilla not? What was this soul? Something only humans had? My head was in turmoil; the fundamental beliefs of my life were falling apart.

Jinka, sensing that something was wrong, nudged my arm. Did God think Jinka was inferior to those louts at school? My thoughts strayed to frogs. They had the same body plan as me – five fingers, five toes and two eyes – but I was warm-blooded and had a four-chambered heart whereas they were cold-blooded and had a three-chambered heart. Yet was I more fit to live in a creek than a frog? Would I survive and it not? No, it would survive, just as the beautifully camouflaged moth that I had been admiring would survive. They were the creatures fit for the bush, not me. Was Heaven just for people, then, and not for Nature? Would I live in Heaven for all eternity without the world I loved? Without Jinka? Where was God in all this? What did the Bible have to say about Nature except that humans had a soul and nothing else did? The heresy of the article was jaw-dropping, but so was its truth.

I slammed my fist onto my desk and probably went close to breaking both. I started sobbing with anger. My mother had lied to me, my church had deceived me, and my school was a fraud! My father was away, so I turned on my poor mother and announced that I didn't believe God existed. She had never heard anyone say anything so dreadful in all her life and she burst into tears.

Next I stormed down to St Alban's, and at first rendered the unfortunate deacon speechless with shock. Then he started pontificating, finally telling me that if I didn't renounce my ways and beg God's forgiveness I would burn in Hell for all eternity. He managed to frighten me, but not into submission.

That deacon thinks he's so special, yet he has no idea what I'm talking about.

The following Monday, Mr Dixon, the school's chaplain and the one teacher I could still call a friend, looked relieved on learning that my request to see him urgently was only about evolution. He sat patiently while, agonisingly slowly, I stammered out my newly acquired belief that humans, along with all other life, had evolved by natural selection and that this had nothing to do with God. This was clearly nothing new to him and he had counter-arguments ready to go.

'It's just a theory,' he said, 'an old theory that comes up every now and again which might apply to animals, but not to humans, who are the children of God.'

When I said that this made me question the very existence of God we both knew I had crossed a line. Mr Dixon moved me on to the headmaster, which was a relief because if it had been Bert Finlay I would have been thrashed again and sent back to class – that brute's cure-all for any problem. The headmaster, however, was a rather pious and thoughtful person. He told me I was under stress and sent me home with a note to my parents requesting that they keep me there for a fortnight.

My mother thought this was one step short of expulsion. Maybe so – I had lost all sense of justice. I'd never felt so alone in my life. There was nobody to talk to. Two wardens from my church beat a hasty retreat when they tried to quote the Bible at me and found I knew it much better than they did. Mum tried her best to reason with me but only ended up crying again. Dad made sure I saw him reading the Bible one evening while he was sitting up in bed. It was upside down: he was pretending. This was the one reassurance I had that I wasn't a complete outcast.

A few months later my mother found me a speech therapist who, incredibly, cured my stutter in a little more than a month by making me repeat tongue-twisters over and over again, faster and faster. In just that short time I learned to speak clearly, even quickly. The psychological pain relief was indescribable. By then I was quite an introvert, but that didn't stop me from talking at any opportunity about the one subject that held me captive – why the Bible was wrong. By the time I left school I doubt there was anybody in my class, if not the whole senior school, who did not believe that humans had evolved from apes and that the God-talk continually showered on us was bullshit. Not a good look for a church school – I savoured that revenge.

I felt sorry for Mr Dixon, but soon discovered there's more to Christianity than the theory of it. Not long after I was old enough to get a driving licence, I bought a tiny 1934 model Austin 7 ute, or at least the remains of one, which I managed to get going. As soon as I felt I could get away with it I drove it to school, something strictly prohibited. That afternoon I chauffeured six kids, all yahooing, waving their school boaters and smoking cigarettes, down the Pacific Highway. It was no easy feat: a couple had to stand on each running-board and two more stood in the ute's tiny back tray, hanging onto what was left of the vehicle's tattered canvas top. When we reached a steep bit of highway just before Pymble railway station I felt the brakes failing, so I swerved off the road and ended up, ute on its side, in someone's front garden. Kids were sprawled everywhere, still brandishing cigarettes and yelling 'Whoopee' and 'Get 'er goin, Charlie.' Gasping for air, I looked up straight into the face of Mr Dixon, who was walking home from the train station. Life was different in those days: he said nothing, perhaps noted that we were all apparently undamaged, and walked on.

I thought I was a goner – cars and smoking were at the top of Barker's crime list – but at school next day nothing happened. That lovely man of God could have finished me off then and there. I realised then that he practised what he preached.

I can't complain about the way my classmates treated me in senior school. I didn't make any close friends at Barker but over the years I did hang out with the alpha group, probably because they were so high up the social ladder they didn't need to shun curious misfits like me. And I did try to be at least a little bit cool, pretending to like what they liked, especially the pop music of the time (unhappily I still can't forget Johnny O'Keefe's 'Shout'), and occasionally I went to their homes for a birthday party or similar.

Then I was invited to a real party, one with girls. And dancing. I could not force myself to go in the front door, so I spent the evening sitting in a dark corner of the garden, only emerging when the coast was clear – of girls – to thank my friend's mother for a great evening.

I started hitchhiking to Collaroy beach after that, partly because being by myself I needn't pretend I was anything other than a nerd. And I could try catching waves. Hitchhiking was always easy for the baby-faced pretty boy I was. Usually I was picked up by a man and sometimes I didn't like the way the conversation headed. A couple of times I demanded to be let out, a hand having strayed onto my leg. Minor incidents, but they made me angry and it occurred to me that girls must put up with this sort of crap all the time.

As my final year of school approached I found myself in an increasingly impossible dilemma. There was plenty of evidence, and not just from exams, that I was indeed not at all bright. One day a prefect put me on lunchtime pool duty because he had something better to do. My job was to enter the names of anybody who broke the rules in the Pool Duty book, and if need be, ban him from the pool – not that anybody would do as I, of all people, ordered. My cousin, a couple of years younger than me, was in the pool that day and probably deliberately did something he shouldn't. I couldn't have cared less but, duty calling, I cautioned him. He did it again, so I entered his name and crime in the book. He came up to see what I'd written and saw that I'd spelt his name Whales instead of Wailes. I was no doubt thinking about humpbacks.

'Don't you know the name of your own grandmother?' he said.

Was I really that stupid? Apparently yes – this sort of thing happened all the time. But other kids didn't think about important things like evolution and religion. They couldn't memorise the eyesight test at the police station, as I did, so that a half-blind classmate could get his learner's permit. And I knew a lot about subjects they didn't. I could make no sense of it all.

Just occasionally I had a chance to fight back. One of Sydney's radio stations had a weekly general knowledge program called *The Quiz Kids*, something I could test myself against. I usually lost against the reigning champion, but did well against most contenders. I asked my mother if I could apply to go on the show.

'Don't be silly,' she said, perhaps imagining the embarrassment of a zero score.

Silly? She had no idea that I remembered just about everything I read. Remembered and wondered about it. I didn't know that yet either; it didn't occur to me that what I wasn't good at was regurgitating facts I was told I must know, especially in an exam room. Most kids will come to believe they can't learn if they're told that often enough, and so my failures in maths, physics and chemistry were self-fulfilling prophecies. When doing an exam in other subjects, I would furiously expound on the first questions only to hear the 'pens down' call before I had started the rest.

The best part of most school days, apart from talking to my mother in the kitchen while shelling peas or mashing spuds, was the reception I unfailingly got from Jinka. He was always waiting for me in the front garden and as soon as he saw me he would charge, as often as not knocking me flat. Life was never bad when he was with me; he was a dog soulmate and my fellow escapee to another world, my own world.

On weekends we sometimes went back to the bush, as we used to when I was small, except that now I took a rope and got Jinka to tow my bike up steep bits of the track, or along the road home when I felt I couldn't make it. I had a light on my bike powered by a generator run from the back tyre; it gave off a dim glow when dark descended. But Jinka had no need of it. I had to steer desperately to keep him in the light as he towed me at full gallop along winding, tree-enclosed paths.

One weekend I persuaded my mother to drive to the beach in our family Morris Minor, a good excuse for her to enjoy a little surfing and me to bring Jinka. There was a good surf running, so we set up our umbrella and I told Jinka to sit and then followed Mum out through the breakers to await a wave. After some time, I was astonished to see a big golden head struggling towards me; Jinka had followed me out. We went back as best we could, my hand gripping his collar, pulling him up when a wave broke over us. What a hell of a time he must have had going out through those waves; he was a good swimmer, but a dog in 8-foot breakers? He was so exhausted he could hardly walk. Again I told him to stay under the umbrella, but to no avail; he waited for a few minutes, then started struggling out after me again. *Bloody dog, why don't you do what you're told?*

As that final school year dragged on I yearned to leave the city and live in the country, in the bush, or maybe on a sheep station like one near Narrandera we visited for a holiday. With my father's occasional tutelage and the use of his tools I became quite a mechanic, servicing and occasionally repairing our Morris Minor or any other car I could lay my hands on. Before I bought my Austin 7 I also built a wreck of a car myself with parts from a local car cemetery, all good training for life on the land. I decided that on leaving school I wanted to go to Hawkesbury Agricultural College, which wasn't far from Sydney, but that meant getting a Commonwealth Scholarship. Having Mum remind me of the sacrifice my family had made in sending me to Barker, there was no way I was going to accept any further support from my parents for education. Hawkesbury was a goal I could aim for, especially when doing homework. I would write out a chemical equation a dozen times, trying to memorise it. A week later I'd get it wrong and do it all again. And all the time my inner self nagged at me, claiming that the dumb kid who couldn't memorise such things was someone the school had made, and that he was not me.

The day of reckoning came with the results of the Leaving Certificate – the public end-of-school exams. I had passed, but with the lowest score possible, so there was no way I was going to get a scholarship. Hardly anyone in my class had done so badly. I felt crushed, and envious of my friends, all of whom had done well. They could go on to university, if that was their choice. It seemed that I was never going to have a life I wanted to live.

Dad sent me off to see a vocational guidance adviser, who looked at my results, heard that I liked animals and mechanical things, and advised me to seek a job as a mechanic in a zoo. My parents saw a bank manager they knew and came home with the happy news that he might take me on as an apprentice.

Mum, Dad, do you have any idea who I am? I'm your son, remember me? I belong in the country. A bank is just about the last place on earth for me. I'd die.

After much gloomy soul-searching it was decided that I would repeat the school year. I buried my pride, such as it was. Barker, surprisingly, agreed to have me back. I was to be exempt from extracurricular activities and biology, which was the only subject I had consistently passed. Instead I was to spend these periods in the library, revising. Dad arranged for me to have private tuition in maths, which was general maths – maths for dummies – but that didn't work out. I thought it strange that my father, with an engineering degree, never helped me with maths himself.

By the time the exams came around again I was more concerned that I'd done no revision for biology during the year than anything else. I needn't have worried about that as I got the highest grade, but my results for the rest of the subjects were exactly the same as the previous year. Goodbye Hawkesbury Agricultural College. I decided to hitchhike west and become a jackaroo, somehow, somewhere.

About a fortnight later Dad produced a letter from the Department of Education informing me that I'd failed to get a

Commonwealth Scholarship but instead offering me aptitude tests – whatever they were – for a second round of applications. I didn't greet that with the slightest enthusiasm.

'Dad, I've tried, and failed. Twice. I've had all the exams I can take. I'm off. I'm going to be a jackaroo.'

'Just one more,' he said. 'You don't have to do any homework for this.'

'I'm not doing it, Dad.'

'Yes you are.'

The tests were held in a large gymnasium crammed with desks and hundreds of kids. I filled in masses of multiple-choice questions, IQ tests and the like, none of which required any knowledge of anything. I was one of the first to finish and leave. That no doubt meant I had failed again, but I didn't care; I was just glad to be out of there.

This is the last exam I'm going to do for as long as I live.

A fortnight later Dad received another letter; it required me to do another round of tests.

'Bugger that, Dad, I'm off. I'm not doing this stuff any more.'

'Yes you *are*.'

'Dad, I'm *not*.'

These tests were in a smaller room with a lot fewer kids. Otherwise it was more of the same.

There followed a particularly empty time for me. I spread a *National Geographic* map of Australia on the floor of my bedroom and pored over it, wondering what it would be like to go to this place or that. But how could I take Jinka with me, wherever I was going? We spent most days in the bush. At least that was something I could enjoy. Every kid I knew was starting out on a new life, doing something he wanted to do. They were a world apart from me.

A fortnight later yet another summons arrived. Dad pushed his military-trained control of me to the limit. 'You're going.'

This time I was ushered into a waiting room in a city office block with just a few other kids. My name was called and I was shown into an office where a stern-faced woman and four men sat behind a long desk. They all faced one chair, which I was asked to sit in. I didn't know what was going on; it was very unnerving.

'We'd like to know what you want to do,' said the stern-faced woman, who was sitting in the middle.

'What? Do you mean to say I have a scholarship?' I was in a state of disbelief.

'Yes you have, but that's not why you're here,' she said impatiently. 'We're offering ten scholarships to applicants who are, as we say, gifted. The scholarship allows you to study for any degree in any Australian university.'

The shock of this brought back my stammer. 'W-what am I s-supposed to be gifted at?'

'Cognitive reasoning,' said one of the men in a deep voice.

'What's that?'

'Students like you are usually exceptionally good at mathematics. In fact – I shouldn't say so but you got the highest test score in the state. This doesn't mean that you know a lot of maths – we didn't test for that, as you know – it means that you're exceptionally good at finding answers to complex questions, any sort of question.'

These people have got the wrong person.

I stood up and told them I'd never passed a maths exam in my life, this was all a big mistake. I thanked them, apologised, and headed for the door.

'Yes, we've been talking to Barker College about you,' said another man to my back.

I returned to my chair dumbstruck.

'So what do you want to do?' the woman asked again.

'I, er, I want to go to Hawkesbury Agricultural College.'

'Young man,' said deep voice, 'people with this scholarship

don't go to agricultural colleges. This scholarship allows you to do anything at any Australian university – law, medicine, vet science, *anything*.'

'Do I have to decide now?' What else could I say?

'No you don't,' said the woman, 'but remember that this is a very rare opportunity. It's part of a study, which will probably only run for one year, to determine how well school results predict university performance. We hope it will help us assess and improve school teaching methods.'

This turn of events reminded me of a neighbour we'd once had, Justice Windeyer, who'd lived two doors up from us in a shambles of a house that nevertheless had an excellent vegie garden with chooks and ducks. I used to help him in his garden, and as his eyesight wasn't good it was me who found the cabbage moths and aphids before they could do much damage. He would show me which insecticides to use, and how to use them so that they didn't harm honeybees. We spent much time discussing his garden, and when he became terminally ill he told my mother that his only regret in dying was that he wouldn't get to see what I did with my life. That comment, when I heard it, was extraordinary to me, and became something of a prop during my senior school years when I had ample cause to wonder the very same thing.

Freedom

It didn't take long for me to work out that there was only one university for me, the University of New England at Armidale, a small city on the high tablelands of eastern New South Wales. At that time this was Australia's only rural university. I would have preferred to be by the sea, but no matter, I just wanted to get away from the city, all cities.

Leaving Sydney – in February 1963 – felt like being released from prison. Good, but not all good, for I might have worn a label: 'Damaged Goods'. I knew virtually no physics and no chemistry. I was unsure how a slide rule worked, hadn't a clue what calculus meant, and was doubtful about the difference between a cosine and a street sign. My spelling was atrocious and my grammar worse. On top of that, I could not play any sort of sport, and worse still, I was withdrawn and lacking in most teenage social skills.

The train journey from Sydney took about eight hours, the track being steep in places. All freshers, as we were called, were met at Armidale railway station and bussed to one of the university's five residences. Mine was Wright College, an assemblage of slightly dilapidated wooden dormitories with a new brick office and dining complex in the middle. My room was on the upper floor of the farthest dormitory. On arrival, I looked up at the stairs in dismay.

How am I ever going to get my suitcase up that? I can't ask for help on my first day.

Surprisingly, I reached the halfway landing still breathing easily and so continued to the top, and that evening I walked half a mile or so to Booloominbah, the Victorian mansion that was once the home of the university's founder and first chancellor, P.A. Wright, and which had ever since been the administrative hub of the university. On entering the massive front doors, there was P.A.'s portrait, his gentle, weather-beaten face smiling out on his world. With a great deal of foresight, I might have said, 'Hello P.A., I'm a guy called Charlie who will thank your university for the rest of his life, although right now I have no idea what I'm doing here. But more importantly, your daughter Judith is about to be one of the people who save the Great Barrier Reef from mining and oil drilling. A real heroine. Well might you smile.'

In reality all I was thinking was that I was walking around with ease; my asthma had completely disappeared. After five years of that debilitating illness, the freedom I felt beggared belief.

My life – the scholarship and now this – had done a complete U-turn. I had been reborn.

Even so, there was a pressing question facing me: what should I study? I decided on psychology in order to find out who I was, why I'd got my scholarship, and to fill in the rest with whatever sciences were on offer. That choice was a problem for the university since psychology was in the arts faculty. Nevertheless, the psychology department welcomed me with open arms, not as a regular student, but as a lab rat. They too wanted to know how a chronic underachiever had been given the rarest scholarship in the land. And so it was agreed – I could do all the psychology I wanted in return for participating in 'tests'. These tests, mostly interviews, continued for eight years, on and off, and probably ended up in someone's PhD thesis. As agreed, I never asked what they revealed, but I suspect most of it was just more unanswered questions. There was much investigation of my relationship with my parents, especially my mother, but chiefly the interviews were about my feelings of deep connection with Nature.

I never did discover why I got the scholarship, but I did find out that I was not an apprentice Einstein. Still, I never failed another exam, not even in chemistry, which in any case was based on the periodic table and quantum mechanics and not remotely like the drivel old Loco Lathum had dished out. In fact I thoroughly enjoyed chemistry and gained a solid grounding in it.

Sometimes I felt bitter about what I had missed out on at school – maths certainly, but a complete absence of music headed the list. I knew classical music existed but had never heard a single note of it. We had a radio in the kitchen that Mum listened to but it was always tuned to 2GB, a commercial station not strong on classical music. We also had a wind-up record-player for 78s, which I loved listening to – mostly wartime songs, Irish folksongs and Negro spirituals that Dad occasionally bought. Barker had two honky-tonk pianos that stood in small cubicles on the school's main concourse, but heaven help any boy who attempted anything more sophisticated than 'Chopsticks' – he would have been branded a pansy and persecuted no end. On my first day at Wright College I heard someone playing a Bach Brandenburg concerto on their record player and I was absolutely transfixed; I had never heard anything so beautiful in my life. I stood riveted, and then quickly went on to discover something about myself that hitherto had been completely hidden – I profoundly loved classical music. It became a part of my being, almost as much as Nature has always been, sinking ever deeper into me, to the core of who I am. Any time I was alone, or when a lecture was boring, I would play a Beethoven symphony, or listen to some opera or maybe a piano sonata. In my head.

A couple of years on, full of enthusiasm and newly acquired knowledge, I gave music appreciation classes at my college – fortunately, as I later discovered, in the absence of any musicians. What a farce; thank god I had enough sense to steer clear of giving sex appreciation classes.

During my first year at Wright I built a record-player, consisting of two large speaker boxes, each with four speakers, including a 12-inch woofer. The rest of the machine was a Telefunken player and a Playmaster 4 amplifier that had come, valves and all, in kit form. With a staggering 30 watts per channel it was the most powerful record-player at Wright, and seemingly the whole university as it was constantly being borrowed for parties. That was okay by me as it meant I could borrow records everywhere in return, and soon I had an excellent mental road map around the colleges of who had what music. Not party music of course, classical.

'Charlie, will you turn that bloody racket down,' a voice would come through the wall, or floor or door. Not everybody liked Mozart.

'Sorry, Rob.'

It wasn't until twenty-seven years later that I bought a new record-player. No one could understand why I put up with such a dreadful old machine. I suppose I'd just got used to it.

Even today I sometimes have to pull over to the side of the road when driving, so captivated am I by music on the radio. It makes me oblivious to all else, an accident waiting to happen.

On my first day at Wright College I met Ric How, who had a room opposite mine. Against all odds we became very close friends, and remained so all through our years at university and on into married life. Ric had more of nature's handouts than just about anybody I knew, and unsurprisingly had been head prefect of his school. He was an ace football player and bright into the bargain. We made an odd pair, Ric and I, with him seeming to turn the head of every girl we passed, and me being afraid that one might look at me.

After the movie *Alfie* came to town my name at Wright became Chalfie for a while, Alfie being a character who had spectacular successes with women. This might sound sarcastic or malicious but it never was. Names were friendly fire, usually subtle, even clever, if at times a trifle uncouth.

None of us worked hard, except when exams threatened. There was just too much else to do. I spent a lot of time eating enormous meals and jogging. With my asthma gone, I revelled in my new-found freedom, jogging for miles every day. As winners of special scholarships go, I didn't make much headway with the more intellectual side of university life. I joined the chess club, not such a good idea as most of its members were academics rather than students, but I only lost one game and, strangely, that annoyed me. I'm a good loser, having had so much practice at school, but chess was one thing I seemed to be good at, at least by the standard of those days. My sort of chess was not the cerebral sort, it was basically pattern recognition, which involved little thinking, just quick comprehension and not making a mistake. I lost that one game because of a mistake – infuriating.

I still had time to spend in the bush, loading my canvas and steel H-frame with whatever food I could scrounge from the college dining hall and then hitching a ride to nowhere in particular. As befitted my age I seldom gave safety a second thought: I never told anyone where I was headed, which wasn't sensible given that my favourite places were off the New England escarpment in very rugged countryside. I missed Jinka on these outings, but even if he'd been with me he would have been too old to cope with that terrain. Unless completely lost, which I was a couple of times, I only stayed out a night or two, sometimes spending much of it in pouring rain, protected only by my father's army wind-jacket and a tiny nylon pocket tent. I always tried to find a quiet place beside a creek to camp and light a fire, if I could. I'd relax, sit with a log for a backrest, and let my mind drift into emptiness. When I came to I'd play some music in my head, perhaps watch an ant drag its dinner to its nest. Music, Nature, and the sound of a creek are an intoxicating mixture, almost to be feared as a drug of addiction.

Darkness comes stealthily in that part of the world. The creek would trickle on but the water would stop sparkling, turn a translucent grey and then get swallowed by the night. I would be enveloped by the dark, except for the friendly glow of my fire. Something might startle me by crashing behind me, but then all would be still except for the monotonous mantra of frogs, the occasional rustle of leaves, and the creek. Drops of rain might make my fire hiss in defiance. I'd scramble to my tent, getting in as my fire was wiped out, and then nothing could be seen. Perhaps I would accidentally touch the tent top, making it leak on my head. I would pull up my wind-jacket. There was nowhere to go and nothing to do. Just absolute solitude. I might brush a bush cockroach from my face, being careful not to hurt it, then fall into a deep dreamless sleep, enshrouded by peace and without a care in the world.

I wasn't always alone in the bush. Sometimes Ric came with me, and I enjoyed his company when we headed off in search of the brush-tailed possums he was studying and the gliding possums I was supposed to be studying. Ric had a favourite field site in the Dorrigo National Park, on a rugged escarpment whose V-shaped valleys were steep and whose trees were among the tallest in the country. I will never forget a time when we were heading for our base camp, a deserted logger's hut, and heard the unmistakable sound of a timber jinker start up. Not just any timber jinker but one I immediately recognised, having taken it out for a short spin along a rough track, the driver yelling instructions at me. It was an ancient Leyland, designed to replace bullock teams, with a massive engine and two gearboxes end to end, giving it thirty forward gears. The old monster was slow, but with railway sleepers for a bumper bar, was almost unstoppable. From deep in our valley, Ric and I stood listening to the engine revving and roaring as the driver double-declutched down the gears before tackling a steep climb. A little further on we listened to him go up through the gears as the track levelled off and he picked up speed. But only briefly; soon he was going back down the gears again in preparation for a steep descent. After him we went, only to find ourselves getting deeper and deeper into an ever steeper valley and increasingly impenetrable bush. Next a chainsaw started up, *rumm, rumm, ruuuum*, followed by the *clop clop clop* of a bushman driving a wedge, before the timber jinker resumed, heading deeper still into the valley. We came to within a few yards of the lyrebird, for that's what it was; we'd had no idea they were such skilled mimics. I could almost tell what gear it was in.

Of the many adventures during my undergraduate years there's one I could have done without. I joined the mountaineering club, which had trips to wild and rugged places that suited me perfectly, especially when I became lean and fit with a good power-to-weight ratio. I wasn't an expert climber by any means but I did love a challenge. The equipment we had was primitive by today's standards, but it sufficed for moderately serious climbs.

One cold blustery afternoon I climbed Dangar Falls, a near-vertical face about 700 feet high. After heavy rain the falls thunder down, a spectacular sight, but during the dry there's just a jagged, black, intimidating wall of rock. It was a climb unlike anything I'd attempted before. When I looked up from the bottom I regretted my decision to tackle it. Twice on the face, my arms and legs shaking with exertion or fear or both, I thought I should go back down, but that looked harder than continuing up. I was on a safety rope, belayed from above by a newcomer to our club who was anchored to a tree. Fortunately, I had given him my father's much loved, indestructible army wind jacket to put on over his woollen jumper, for just as I swung myself over the final ledge with a cry of victory, I slipped. I only remember revolving over and over, hitting rocks and thinking my back would break from the force of the safety rope. I reached the bottom of the falls with a hell of a thump, after bouncing off rocks most of the way.

I regained consciousness in a haze of pain and concussion, then saw the others racing down a track to the bottom of the falls to reach me. Immensely relieved to find me still in one piece, they were full of praise for my belay partner, who'd acted with great speed and courage to break my fall. In the process the safety rope had burnt through my father's wind jacket, then his own pullover, and started scoring his back. In a state of shock, I insisted I was fine and could walk out unaided, but when most of my memory of the fall came back several hours later I couldn't recall if I'd thanked my protector for saving my life. We agreed to keep quiet about the incident in case the university closed the mountaineering club down.

One summer holiday in 1967, at the end of my fourth (honours) year, I spent three glorious weeks camping on Hinchinbrook Island, in northern Queensland. There were ten of us, all men of course, including my two closest friends, Rick Smyth, by then studying dentistry in Sydney, and Ric How.

Today the island has a major resort, and permits are required even for a day visit, but no one lived on Hinchinbrook then. In fact were it not for the occasional distant fishing boat far out to sea, we could imagine that we were the only people on Earth. The nights were a world apart. With no city lights and no dust in the air, the heavens sparkled with a brilliance I'd never seen before. Working out star constellations without a guidebook became a nightly challenge, and most nights after dusk satellites passed overhead, tiny specks like fast-moving stars doing what, we could hardly guess. Peace and quiet surrounded us. Peace in a timeless place.

We camped at Zoe Bay and climbed the island's highest mountain, 3670 feet, the second-highest in Queensland. I led the climbing, taking a rope up and tying it to trees in case the others needed it. It wasn't a difficult scramble but I kept moving to the left, where the scrub seemed thinnest, so it was my fault that we found ourselves on the wrong – steepest – side of the mountain as dark descended. There was nothing for it but to spend a precarious night on the mountainside, on platforms made of whatever scrub branches we could weave together. My platform collapsed in the night and at dawn I found myself about 5 yards down the slope without having woken. We reached the top that morning but were enclosed in fog. Some of our number, fed up with my bushwalking enthusiasm, went back down (by the route we should have come up) but a few of us stayed the next night, hoping for a view the following day.

As dusk settled, large white-tailed native rats turned up. They had no fear of us whatsoever and wanted to fight us for our dinner. We gave them bits of what we had, but that didn't satisfy them so they bit holes in our li-los in the night. No matter, next morning we were rewarded with a spectacular view of the whole of Hinchinbrook, the Palm Islands, and a hundred miles of coast. I never dreamt that one day I would visit all we saw many times, eventually coming to know most of the wildlife that lived there.

We had paid a fisherman to take us over and pick us up two weeks later, weather permitting. The weather was perfect, but a third week passed before he turned up, unconcerned – a genuine laidback north-Queenslander. I was getting a little worried we'd be stuck there until we starved, but we got by with what we had or could catch.

Apart from a couple of big trips to Central Australia with the university's Exploration Society, some of my best student days were spent travelling on my own to one beautiful part of the country after another. After my first year I did a mind-blowing trip to Heron Island in the southern Great Barrier Reef, hitch-hiking up and back and staying at the research station there, a delightfully dilapidated building used mainly for teaching. I don't doubt that this place weaves its magic on all who go there; it certainly did on me. It was the most captivating place I'd ever laid eyes on, its reefs bursting with life I'd never imagined. I wanted to live there forever. Above all, I wanted to learn to scuba dive.

I still wonder about one incident that occurred there. I met two French underwater photographers who wanted to film manta rays and I knew just where to find them: a flat sandy place, about 80 feet deep in the channel between Heron Island and Wistari Reef, where two mantas kept swimming in endless circles, or figures-of-eight. All the divers needed to do was get to the right place and wait for the mantas to come to them. Off we went, me snorkelling on the surface and the photographers following on scuba, and sure enough, along came the mantas. I wanted to see one of these magnificent creatures close up myself, so judging my time carefully, down I went. It was thrilling; one of the mantas, about ten feet across, brushed right over the top of me. Then I begged a breath of air from one of the photographers so I could stay down a little longer. He should have known better than to give it to someone who'd never used scuba, as free divers hold their breath as long as they can, and expanding air on ascent is dangerous. But I must have breathed out unknowingly.

At Wright, there was a culture of being men in a man's college. Unhappily, the sum of my undergraduate social life with women was getting to know some of Ric's female friends during casual outings to football matches and the like. The one exception was the compulsory annual college ball, to which I asked the same lovely but safe girl every year. Poor Chris, I hope she found someone interesting to talk to; it surely wasn't me.

I spent a lot of time and money drinking in pubs, as most students did. No women were allowed in a front bar in those days, so our behaviour was awful, worse than I'm going to admit to here. All Armidale's pubs were in easy reach of our college because there was only one road between the university and town and most drivers picked up students. Coming back to college in a student's car after ten o'clock closing was sometimes an interesting experience, but the police were tolerant and if there was an accident, being drunk was always a good excuse. One evening a mate and I hitched a ride at closing time by jumping onto the bonnet of the first car to pass, which happened to be a police car. The two of us spent the night in the police watch house. We sang bawdy songs until the sergeant threatened to take our sleeping mats away.

Pranks were a part of everyday life for us, often hilarious and just as frequently infantile. Some are unthinkable today, like the time we bade farewell to a popular college tutor at the railway station. We managed to wrap his entire carriage in toilet paper and then unhook the engine. The stationmaster, on arrival, took the wrapping well enough, but when the engine moved off without the train, the driver wasn't at all impressed. We calmed him down by agreeing to hook his train back on as he backed up. It seems that a lot of the fun has gone out of student life these days. That's a pity, for it's harmless enough and part of growing up, of feeling free.

As a university student, 1967.

Delayed metamorphosis

Perhaps I might have gone to the Amazon or some such place after graduating had fate not intervened in the form of a very attractive freshette called Kirsty Mackenzie. By now I was a different person to the chronic underachiever who'd first come to Armidale. I had become very much part of my all-male college, enjoying the good life, but I was still absurdly shy as far as women were concerned. In 1966 I found myself elected to the unpaid job of secretary of my college, one of my duties being to invite the junior common room committees of other colleges to an annual formal dinner.

These were normally stressful affairs for me, but that year Kirsty was among those who attended. She was the first-year representative of the newly formed Duval College, just across the road from Wright, and she seemed interested in my accounts of bushwhacking with Ric. I plied her, and myself of course, with what I thought was red wine but was in fact sherry. I didn't know the difference. So, fortified with excess Dutch courage, I staggered up to my room and brought down my tiny pet pygmy possum to show her. She was very impressed. That did it. The next day I dredged up enough of my own courage to see her again. Then again, and again. My female-relationship drought of twenty-one years had broken.

Then came another quirk of fate: I 'won the lottery'. All able-bodied Australian men born on the sixteenth of February, as I was, or on other birthdays selected by a lottery, were conscripted for military service in Vietnam. Like many other students, I'd been demonstrating against the war, once helping to block all traffic to the university on a bridge below the main campus. Maybe I could avoid conscription by claiming I had moral objections to the war? Perhaps my part in the demonstration on the bridge would suffice? If that didn't work, maybe I could go to prison instead of Vietnam? In all honesty, my conscientious objection was not only to conscription, but to being separated from Kirsty. I decided on prison rather than war, not exactly a romantic choice.

Next trip home I told my mother about being conscripted and that I was refusing to go.

'Don't you *ever* tell your father, it would break his heart,' she replied.

Wow. I'd expected that reaction from Dad, but not my wonderful, all-loving, all-caring mother. Wartime white feathers appeared before my eyes.

'Dad,' I said, in full confrontation mode when he got home a short time later, 'I've been conscripted for Vietnam and I'm not going.'

'Of course you're not,' my father said. 'The whole thing's a ghastly political farce.'

Two months later I received a summons for a compulsory medical examination at the Armidale hospital. Neither Vietnam nor prison seemed much of a substitute for Kirsty; I had to get out of both, I had to fail that medical. This was quite a challenge as I was medically fit for just about anything imaginable. I contemplated breaking an arm or leg somehow, perhaps not a simple task considering the innumerable injuries I'd survived during my bush-going ventures, not least that fall down Dangar Falls which should have finished me off for good. A better idea was to get bronchitis and claim, with medical records that hopefully still existed from my school days, that I suffered from asthma. The medical was only a week away: how could I get bronchitis in a week when I hadn't had it in years? I went for hard runs along country roads, getting hot and sweaty, then dived into my college kitchen's freezer room, sitting there in my jocks until I could stand it no longer. I did this every day, much to the amusement of the women in the kitchen.

Of course I didn't even catch a cold. On the day of the medical I went for a last run along a different road in the hope that something would happen. It did. On the way back I saw an old haystack in a cow paddock, and jumping the fence and ignoring a chorus of moos I plunged in, burying myself in half-rotted hay. After a few minutes I had an asthma attack the likes of which I'd never experienced in my life. I staggered back to the road thinking my end was nigh, when along came a kindly motorist who delivered me to my college. My mates thought it a great joke. They took me to the medical and someone, I don't know who, almost carried me to the appointed doctor. My eyes were bulging and I could hardly breathe. I told him I was always like this, sad to say. The doctor took one glance at me and shouted, 'Get him outta here.'

Goodbye Vietnam, hello Kirsty . . .

Kirsty came from a historic cattle and sheep station called Stonehenge about 100 kilometres north of Armidale. The house was in a state of moderate disrepair but, surrounded by 2 hectares of wild garden, was endearingly beautiful – a home to be treasured, and Kirsty certainly did that. Needless to say, I became a regular visitor, always made welcome by her parents. I loved exploring the property with her on horseback, me on Kirsty's wily old ex-racehorse who didn't hesitate to jump anything. I just hung on for dear life and had a couple of rather nasty high-speed crashes. Much safer to build a fish pond, which I did among some granite boulders in a corner of the garden.

Being with Kirsty put an end to my solo bushwalks. I didn't reflect on that much at the time because a friend of Ric's had bought a mould for making fibreglass Canadian canoes, decked in and with a cockpit at each end. This allowed for bush ventures with a difference. Kirsty and I made an expert pair, she in the bow and I in the stern. We'd have only a crude map of where we were going, if any at all, and there were heart-stopping moments when, looking ahead, we saw that our stream had developed a horizon. That meant a rapid or a waterfall lay ahead. As we got closer we'd hear the roar of the water, and sometimes we shot over small falls that caused our canoe to plunge underwater before resurfacing to carry us on. We smashed a couple of canoes this way, on rivers running from the New England tablelands to the coastal floodplains. It was incredibly exciting, but, I'll admit, rather dangerous.

Forty years on I made a pilgrimage back to Stonehenge. It had been resold a couple of times and fallen on hard times, but by the time I returned it had become a wealthy man's mansion. The owner was delighted to show me around, especially the garden, which was his pride and joy and which I, a lover of beautiful places, found stunning.

'There's a really pretty little rock pool over there,' he said, pointing to a distant corner as we walked around. 'Nestled among a crop of granite boulders and full of orange lilies.'

'I know,' I said. 'I built it.' That took him by surprise.

The following year, honours graduation over with, it vaguely occurred to me that I should start looking for a real job. But what? Fortune, the wildcard of my life, intervened before my job search began. The zoology department needed an assistant for Hal Heatwole, a recently appointed associate professor, and were after an outdoors type – me. My job, among other things, was to catch reptiles, keep them in a newly built snake house, and assist with experiments on them. At the same time I could enrol for a master's degree. I can't say I particularly enjoyed the work, or my own project on lizards, but I loved going on trips with Hal to inland northern Australia searching for snakes.

One day four of us were in a rainforest when we came across an amethystine python, Australia's biggest snake. It was nearly 6 metres long, with a head like a dog's. I wanted to see how hard it could squeeze, so I asked the others to wind it around me, one keeping a hold of the head – no easy task – and another managing its tail. After a little struggling, three loops were around me. Then the snake started constricting. It crushed the air out of my lungs and I felt my ribs caving in. I had no hope of breathing and certainly couldn't speak. At that point the others urgently unwound the snake; I was lucky they still could. A daft experiment, not recommended, but I confess I'm a little odd about dangerous animals and dangerous places – they seem to attract me, especially when I'm alone. A dysfunctional sense of self-preservation perhaps? Maybe, but it's more that I resent unnecessary fears getting between me and Nature. If I had not overcome my initial intuitive fear of dangerous animals (like funnel-web spiders) or dangerous places (like the outer edge of Long Reef) my life would be the poorer for it, just as people who are afraid of the dark suffer a sad affliction. It's a personal thing, although it sometimes encourages me to make others confront their phobias, like when a television cameraman turned up to photograph some of our snakes and wanted me to hold one for him. I picked up a tiger snake and drooped it around my shoulders. Then another and another, until I had six crawling over me. The cameraman backed away as far as he could go. I wasn't being brave or silly, I was merely showing him that it was cold enough to make the snakes placid, and harmless if treated gently. I knew what I was doing. But I don't think this demonstration succeeded, for the poor guy fled just as soon as he could.

Back in the lab, my job had had a seriously doubtful start. Hal needed to collect salt from the salt-excreting glands of sea snakes, which are situated inside their nostrils. How? Hal considered this for a long time and decided that a condom, placed over the nose, would do it. Of course it was my job to get the condom, so I drove into town and paced up and down outside a pharmacy until there were no customers inside. I walked in, only to be riveted to the spot by a rather attractive shop assistant smiling at me. My shyness of old came back with a vengeance. She asked if she could help.

'Ah, no – yes. I, er, well, I think I need some condoms.'

'In the corner there.' She pointed, her smile unchanged.

I bought a packet, pocketed the receipt and fled. Hal didn't notice the trauma I'd been through; he just got on with the job. The first condom tore. He tried another – that nearly did it. Then another – that worked. Job over. This was fortunate as I'd purchased a pack of three.

A couple of days later Hal decided to use this method for the whole study, and he needed a lot more condoms. I drove back to the pharmacy and there she was again.

'Can I help you?' she asked, in what seemed to me to be an ambush.

I mumbled out my request, looking anywhere but at her, and she informed me that Durex came in packets of tens and hundreds. I bought a hundred.

But there were several snakes and the condoms had to be changed twice daily. A week later Hal told me to get more. He had no idea what he was asking of me.

Back to the pharmacy I went and, oh no, there she was again.

'Really?' she said when I muttered my order. She was trying not to grin as she took my twenty-dollar note. 'That should keep you going for a while.'

I mumbled that they were for sea snakes.

'Really?' she said again. 'I didn't know sea snakes used them. See you next week then?'

I didn't wait for a receipt.

Horrible woman.

One evening, on coming back from a long trip, I went straight to the snake house, which was situated on a parapet between the zoology and botany departments. I immediately saw that the heating had failed, and then realised that the technician who was supposed to look after the snakes hadn't bothered to check on them at all.

I was relieved to see my beautiful amethystine python – the one that had been wrapped around me – moving, until I realised it wasn't she that caused the movement. I flung open the door to find that the rats I'd left for her food had eaten their way into her body – she was too cold to defend herself – and were now eating her from the inside while she was still alive. There followed horror upon horror as I opened one compartment door after another. More snakes were dead. Dead rats were half decomposed. Those snakes still alive were moribund – yellow-bellied blacks, tigers, a death adder, a king brown and some copperheads among them.

Boiling with anger, I forgot to close the doors. I just stormed back to my room, there to spend half the night composing a speech about animal welfare. Next morning I went in to work late, not relishing the task ahead. However, instead of having to blurt out my speech I found myself cordially greeted by some zoology staff and two botany professors. The latter were beside themselves: the snakes, warmed by the morning sun, had found their way into the greenhouse, as well as up some scaffolding that builders were using to make repairs. The builders had walked off by the time I arrived and the whole botany wing had been evacuated by order of the vice-chancellor. I was told that louts had opened all the snake-house doors, and would I please oblige by 'getting rid' of the snakes. I said I would catch them, and did. I spent the day taking them to places where I hoped each might survive. The snake house was never used again.

In 1968 I graduated again, this time with a master's degree in herpetology, and then wrote my first journal paper, on the reproductive cycle of a water lizard. The only problem with that was the name I gave the editor: Charlie Veron. He told me I couldn't use a nickname. What to do about that? I couldn't call myself John, so I used my initials, J.E.N. It sounded formal, but with no alternative this became the name on most of my publications.

With another year's work I could have changed that master's to a PhD, but I felt there was no point in that because my project seemed to be on a path going nowhere particularly interesting.

In May that year Kirsty and I were married. The wedding took place in the Presbyterian Church of Glen Innes and the reception was at Stonehenge. It was a very grand affair, with the whole MacKenzie clan and all their friends. We knew it was Kirsty's father's swan song as he had a terminal heart condition, so the wedding was as special for him as for us. I hadn't been inside a church since I was thirteen, so that part was a bit of a struggle but my friends, who were just as religious as me, took it in their stride and drank merrily on.

We went to Heron Island for our honeymoon – that's what research stations are for, after all – and on our return I was offered the post of teaching fellow by Professor O'Farrell, the head of the zoology department, presumably still grateful for my deliverance of his department from the dreaded snakes.

Scuba

Soon after I started my new job, a group of us decided to do something adventurous, like parachuting. This sounded fun and I was all for it, but one short phone call put that way out of our financial reach. What about scuba diving? In those days there were no scuba clubs and I was the only one who had met anybody who'd actually been scuba diving. I phoned the CSIRO's Division of Oceanography at Cronulla, just south of Sydney, and arranged, when next visiting my family, to go for a dive under the auspices of someone who called himself the diving officer.

True to his word, this kindly man rigged me up with scuba gear and helped me, waddling along duck-fashion with fins and all, to a breakwater, then down over the rocks to the water's edge.

'Remember,' he said, 'don't come up too quickly, and don't hold your breath.' He ambled off.

Following this comprehensive training course (*Isn't there more to scuba diving than this?*) the dive itself was thrilling. I could see only about five feet ahead, mostly rock and mud, but the feeling of being weightless in a hidden world was exciting, if a little frightening. Like most beginners, I kept imagining sharks circulating just out of sight, waiting to eat me. I stayed down until I was freezing cold and then surfaced ever so slowly.

That was the sum of my diving training for over a year, and none of my fun-seeking buddies had more. No matter, we picked up snippets as we went. It wasn't until 1970 that some of us obtained a formal diving qualification from a new diving school on the coast. That had nothing to do with learning to dive, we just needed a ticket to get tanks filled at dive shops.

In the meantime, encouraged by some interesting dives on rocky headlands, we decided to form a scuba club. The zoology department had a marine station at Arrawarra Headland, north of Coffs Harbour, only a few hours' drive from Armidale. The station was a large fibro shed and was free, meaning we knew where the key was hidden. What's more, we could borrow tanks from someone we'd met in Coffs.

Diving on the coast was all very well, but the Solitary Islands, visible from Arrawarra Headland, kept beckoning. How could we get there? The zoology department had a 14-foot aluminium boat with a small outboard motor that we could borrow – well, we didn't feel the need to burden the department with paperwork, and the boat didn't seem to be used for anything else. We discovered that as many as six of us could cram aboard without obvious signs of the boat sinking, and if the seas were calm enough we could easily make the mile-and-a-half journey to South West Solitary Island, which had a slightly protected embayment on the western side where we could anchor.

It was there that we found corals growing on rocks. As far as we could ascertain, corals had never been recorded anywhere along the New South Wales coast, so this was quite a discovery. There weren't just a few corals, there were acres of them, in all manner of different shapes, sizes and colours, intermingled with giant anemones, clown fish, and masses of other marine life we thought only occurred on coral reefs. The place didn't look anything like the reefs I'd seen at Heron Island; it seemed unique, while just as astonishingly beautiful.

We decided to make a survey of the marine life there and use it as evidence to have the place declared a marine park. This was a groundbreaking idea as there were no marine parks in Australia at the time – not even for the Great Barrier Reef – or anywhere else as far as we knew. The survey didn't quite work out, although the idea of a marine park did, twenty years later. None of us had studied marine life other than as part of general zoology, but undaunted we each adopted a group of organisms to survey. I was most interested in the corals but they were Ric's job. I took on worms, memories of my childhood collections coming back. Despite our attempt to devise some rules, our diving practices weren't the best, to put it mildly, and Ric kept getting headaches. So I took over corals from him and that's how my lifelong interest in them was born.

We did some great diving and some rather dangerous diving at the Solitaries, and looking back it's a wonder we all lived through it. Sometimes we got John Rotar, a newly acquired friend, to take us out on his big boat, a hulk he had rebuilt after it was wrecked on the Coffs Harbour breakwater, and when we did we brought newcomers along for the trip. We camped at John's place or on a beach at Coffs, where we'd spend the night around a fire swigging wine. (Diving does indeed cure a hangover, in case you're wondering.)

On one such trip I took a novice for her first dive. We were at Split Solitary Island, which was literally split in half by a chasm that extended deep underwater. We swam through the split easily enough, but on our return I turned into another chasm by mistake. This became progressively narrower while the wave surge became increasingly stronger. My companion, being smaller than me, went deep, where the crevice fortunately widened. I didn't know this, so I had no option but to follow down after her in the hope I could somehow grab a fin and drag her back. I became jammed between the barnacle-encrusted walls of the crevice and ended up taking off my tank to try to go deeper, but only got further stuck. In a state of growing alarm, I battled my way straight up to the surface, scrambled over some rocks where the two halves of the island met, and dived down the other side, hoping to get to her that way. And there she was, happily swimming on without a care in the world. My legs were covered in bleeding scratches and my old wetsuit top was in tatters. I'll admit the episode gave me quite a fright.

As we had no money we scrounged our equipment as best we could, and when there was none to scrounge we improvised or did without. I made my own depth gauge out of a piece of glass tubing, closed at one end and marked with notches where a bubble, responding to the doubling water pressure, was supposed to show depths of roughly 30 and 60 feet. Hardly a substitute for the diving computer I use today, but it wasn't as if I took it too seriously. We didn't have watches, or contents gauges on our tanks, so I would stay on the bottom until I could feel my tank running out, then start for the surface, perhaps fifty feet above, drawing in the last breaths of air as I went.

On one particular dive the sea was calm when I went down but when I came back up, lugging a heavy bag of corals, a sou'easter had arrived and the water had turned rough. I had no snorkel, nor any sort of buoyancy vest. The problem lay in getting a breath of air while struggling with the bag. The collection was important to me and I refused to let it go, rough sea or no. After a few breaths of more water than air I decided that drastic measures were called for. There was nothing for it but to sacrifice my weight belt, an ordinary leather-and-buckle type borrowed from my jeans and threaded through blocks of lead. I then discovered that the buckle had moved behind me and I was unable to undo it with one hand. Waves pushed me under again and again.

I'm not letting go of this bag, no matter what.

Struggling to the surface once more I glimpsed our boat anchored some distance away and waved as best I could. The bastards just waved back. Then I heard a voice casually ask, 'Can I carry something for you, Charlie?' It was Terry Done, one of our club members, and he took the bag just as I felt that the waves were closing over me for good. I was exhausted.

On another trip one of the elderly professors of zoology, John

Le Gay Brereton, wanted to come with us. I was still thinking that our borrowing of the departmental boat was a secret, but he'd found out about it somehow. No matter.

The waves close to Arrawarra Headland were usually hazardous to negotiate in our boat, especially with a full load of divers and gear, so our practice was for me to drive the boat while the others swam out beyond the breakers before climbing on board. If a wave looked threatening I'd make a U-turn and a quick run back towards the shore to try again. This time, with the professor in the bow, the boat was sluggish to turn. Up we went, side-on to the wave, before crashing over, upside down. The professor, now under the boat, had to be rescued and dragged ashore, coughing and spluttering. It took all day to get the outboard going again, by which time the wind had come up and the sea had turned nasty. We returned to Armidale empty-handed.

I thought John would have a fit about what we were doing and the way we did it, but much to our delight, when we delivered him home he announced that he hadn't enjoyed himself so much in years. John had been a pilot during the war and afterwards owned his own plane. He was very unlike the other zoology professors.

One of my last trips to the Solitary Islands at that time was a much worse near-disaster. I took three divers to one of the outer islands – an 11-mile journey – and was returning to Arrawarra to collect the rest when another sou'easter hit. The sea got very rough; there was no way I'd be able to negotiate the worsening weather and collect my buddies. Worse still, they wouldn't be able to get up the steep cliffs of the island, they'd be stuck in the water. Back at Arrawarra, I heard that a surf carnival due to be held just south of the headland had been cancelled because of the conditions, but a Mr Hamilton, builder of HamiltonJet boats, who had intended to demonstrate his surf rescue boat at the carnival, was unfazed when I told him about my divers.

'Hop aboard and hang on,' he yelled.

By now the wind was a screaming gale, but off we went, straight into the face of a ten-foot breaker, me astride the boat's roaring V8 engine and clinging on for dear life. I thought my end had come. We crashed through the first enormous wave, shooting high into the air and coming down with a bone-crunching slam right in front of another 10-footer. Then Hamilton — was he completely mad? — turned his boat side on and we disappeared into the curl of the break, the boat almost overturning before he flipped it out as if it were a surfboard. Demonstration over, he charged off to rescue the beleaguered divers, leaping from wave to wave at horrific speed. We returned relatively slowly, at a mere 20 knots or so, with three cold and weary divers aboard. Of course they were happy to be rescued but I, fit though I was, could hardly straighten my back for a fortnight.

In May 2016, nearly fifty years on, I made a nostalgic return visit to celebrate the twenty-fifth anniversary of the Solitary Islands Marine Park and to do some diving. After checking out the corals at the beginning of each dive I spent a little time just swimming around, enjoying the memories that came flooding back. Nothing had changed, the corals were just as I remembered them, except for the number of species I now recognised — more than double those we originally recorded. I saw where we'd made mistakes, but couldn't help thinking that our records weren't too bad for that first attempt so long ago.

At the Solitary Islands in 1971. From left: Terry Done, Len Zell, Ric How.

John Rotar

Serendipity

The University of New England's zoology department in the late 1960s was pretty decadent, but as it turned out this suited me just fine. With many of the teaching staff more interested in golf or the daily newspaper than giving lectures, I could teach one course after another on their behalf, and thus, over the next few years, taught almost every course the department offered. I loved every facet of zoology.

'Charlie,' a professor or lecturer might say at morning tea, 'I'm particularly busy next term and wonder if you'd mind taking my second-year invertebrate anatomy for me?' Or cell biology, or embryology, or vertebrate anatomy, or whatever.

'I'd love to,' I would always say with genuine enthusiasm, knowing full well that the professor would be off holidaying. Yet again.

I found teaching new and different courses a real joy, never giving the same lecture twice. I especially enjoyed teaching mature-age students, for they were always more committed than raw school leavers, particularly the Catholic priests and nuns who did courses during university holidays. I cracked a lot of jokes and so connected well with most of them, many of whom weren't much older than me. Sometimes I gave lectures in the morning outside under a tree – 'Don't take notes, just listen' – while lab work, which usually involved dissections, was inside during the afternoons.

It was a condition of my job that I enrol for a PhD, not a hardship at all except for the need to find one interesting topic out of so many options. Then, early one cold and misty morning I was out jogging when I noticed that some dragonflies, which were normally bright blue, were a dark grey colour. I watched them over several more mornings and saw that as the early fog lifted they went through some interesting behaviour, orientating themselves at right angles to the rising sun, even when flying. At that time they were all grey, but as the sunlight increased they started turning blue.

I mentioned this in passing to Professor O'Farrell's secretary, Mrs Slade, an ancient old thing who fiercely guarded her lord and master's office.

'You must tell the professor about this', she commanded. 'He'll be most interested.'

Professor O'Farrell (O'Barrell to his back) was a man of vast bulk and domineering personality who generally terrorised the entire staff. He had never supervised a student in his life, but when he heard about what I'd seen he announced that this would be my PhD project, and that he would supervise it personally. I had severe doubts about this, especially the supervision bit – I knew I was unsupervisable. I referred the matter to Mrs Slade, who after a moment's deliberation whispered that the professor kept most of the department's research budget for himself but had no idea what to spend it on. That did it; I became an entomologist then and there.

I did most of my field work at Uralla Lagoon, not far from Armidale, a place I'd enjoyed for years because it was full of birdlife and all manner of aquatic creatures. At first light I recorded the behaviour of the dragonflies and measured their temperature with a thermistor I'd made from gold wire that was thinner than a spider's web. I then secured a perspex chamber inside an aquarium full of water so it could be both temperature- and light-controlled, and photographed changes in the colour of the dragonflies in the chamber with a timelapse movie camera, bought on the professor's tab of course. I found the nerve ganglia that controlled the colour change and used the university's new electron microscope, a complex monster of a machine that few people bothered mastering, to reveal anatomical details of the colour-changing cells. Then, with the help of a biochemistry student I knew, we – well, mostly he, to be honest – identified the colour-changing granules involved.

All this, as it turned out, was just for starters. I could change the colour of a dragonfly's eyes, or abdomen, or tiny pieces of its integument (skin) with different combinations of light, temperature and ganglion extract. Imagine if you can a dragonfly with one eye brown and the other blue, and body segments of different colours, then have all these parts change colour in unison, or independently, with changes of heat, light or ganglion extract. It was a fabulous project: I could have an idea one day and an exciting result the next. I found out how and why dragonfly eyes adapt to all manner of experimental variables – heat and light, transplanted ganglia and then neurosecretions. It was all great fun, but it took me more than a year to find the common ground that connected the behaviour and temperature of dragonflies in the field with the many kinds of experiments I did with them in the lab. In effect, I had four separate angles to the project, no single one of which could be linked to any of the others, or make a thesis on its own.

Meanwhile on the home front, Kirsty and I were living in a little old house on the northern side of Armidale which, we later discovered, was the original country homestead of the area. It was homely and welcoming, if rather run-down. By then Kirsty was in the final year of her arts degree and our life together was turning into one long, never to be forgotten honeymoon. And it was about to get a lot better.

Fiona, or Noni as she was always called, was born on 29 April 1970. I'd always loved the company of children, but a new baby was something way beyond my experience. None of my friends or family had babies at that time; in fact I don't think I'd ever had anything to do with one until then. I wanted none of the ways of my father, to whom babies were a woman's business; I was determined to be an equal parent with Kirsty in all ways possible.

The sister at the Armidale Hospital maternity ward was adamantly against me being present at the birth, but I won that battle with the aid of the attending doctor and so helped with the delivery. Noni was my baby as well, right from the beginning. She was beautiful, with none of the wrinkles newborns usually have, and it was soon apparent that she was easy to look after. I wove a cane basket to carry her in, and if we went to a friend's place we could leave the basket anywhere – Noni slept through anything. When she grew too big for the basket, I cut a hole in one end for her feet. No doubt I'm not the first father to have pondered how weird it is for a little baby to become so important so quickly.

In May that same year, Jinka died, aged sixteen. Jan had written to me that he was failing, saying she was going to have him put down. She couldn't bring herself to phone to tell me, she wrote, and her letter arrived too late. No matter what, I would have driven straight to Sydney to be with him on his last journey.

Noni's birth and Jinka's death both happened during a time when my PhD work consisted of four parts that I couldn't connect. It was very discouraging, but then Kirsty found an interesting advertisement in a newspaper. It called for expressions of interest in post-doc work on corals at James Cook University in Townsville. The only condition, apart from having a PhD, was that it had to be based on field work using scuba. It wasn't stated at the time, but this requirement came from moves to save the Great Barrier Reef from both oil drilling and the ravages of coral-eating crown-of-thorns starfish, threats that had made headlines in one news bulletin after another. Also unknown to me then was a demand that an Australian rather than a foreigner be on hand when a coral expert was required to be a witness at government inquiries about The Reef. Nor did I know that the newly formed James Cook University had proposed building a research institute for the Great Barrier Reef.

Although I still knew little about corals, I wrote a brief note expressing interest, vaguely wondering if a post-doc couldn't be turned into a pre-doc. Or maybe a not-a-doc-at-all sort of doc . . . As it turned out, despite the position being readvertised three times, mine was the only expression of interest the university received. It seems that scuba diving at that time was still a thrillseeker's sport, not to be taken seriously by a scholar.

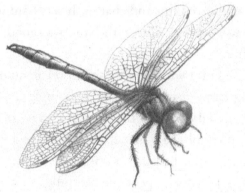

Dragonflies are one of Nature's great triumphs, dating back more than 200 million years.

With Noni filling our lives with endless love, fun and interest, Kirsty and I were keen to have another baby. Ruari was born on 26 August 1971. My mother, who had become close friends with Kirsty and absolutely smitten with Noni, came up to help out. I remember bursting in the front door of our little house shouting, 'It's a boy!' He looked just like me, at least I thought so. I was a bit concerned when he was put into a humidicrib 'for observation', before I left the hospital, but he looked perfect nonetheless, just as Noni had.

However, he wasn't perfect; his birth had been normal but he was a little blue around the mouth. The following morning, I arrived at the hospital as dawn was breaking and was shocked to see that he had an oxygen mask on. I just looked at him, desperate to give him a hug and tell him he was okay. Kirsty was in tears.

Later that day he was flown by air ambulance to Sydney's Prince of Wales children's hospital and there he died, two days later.

Ruari's death left me simply bewildered, and feelings of self-pity flooded over me. How could this have happened to such a perfect baby? To *our* baby? It was my job to look after Kirsty, who was devastated, but it was Noni who did that best. She was always cheerful, full of life and fun and chatter. It was hard to stay grieving when she was with us. Life with Noni was good; we just had to get on with it.

As time moved on I wrote my thesis, all the components of my project having come together to make a single beautiful story. Professor O'Farrell didn't have the faintest idea what I was doing, but someone, probably Mrs Slade, must have told him that I'd discovered something, because he summoned me to his office one afternoon – he always spent the mornings in his bath – and demanded that I go to Canberra to give a paper at the Fourteenth International Congress of Entomology. This was not at all to my liking but I did as I was told. Everybody did what O'Farrell told them to. I took a train to Sydney, then another on to Canberra and there, armed with some colour slides and a page of notes, I umm'ed and err'ed my way through my first paper at a scientific meeting. It turned out to be quite a day: I won the prize for the best student presentation and was offered four post-docs on the spot. Then the editor of *Journal of Insect Physiology* asked me where I had published my work.

I hadn't. I hadn't even given the matter a moment's thought.

'Do you realise, young man, that if you don't publish your work immediately someone will steal it from you?'

This came as a shock: it hadn't occurred to me that anybody might publish my findings as their own, and I had no idea what to do about it, particularly since students were forbidden to publish their work before their degree was conferred. How times have changed. Despite this, the editor suggested I write it up as best

I could and send it, and my colour slides, to him to 'fix up' and publish in his journal.

Back home I did as he said, and that wonderful man did as he promised.[1] Things were looking up. I had four post-docs to choose from, one of which was in Canada and came with a large research budget to work on hormonal control of migratory locust outbreaks, which were decimating crops around the tropical world. Yet something about that offer wasn't quite right. Would this post-doc give me the freedom to do as I wanted or would I just be a small cog in a big machine, unable to follow my own nose? Worse, the journal editor's words kept repeating in my head; I imagined I'd be working with the sort of people who stole others' work.

On hearing this, O'Barrell told me to go away and grow up. Nevertheless, I kept delaying, until I finally took the plunge and accepted the offer of the post-doc in Canada.

A few days later I heard the postman's whistle and took delivery of a telegram from James Cook University offering me the post-doc they had long been advertising. I wrote a polite 'no thankyou' for a job I had dreamt of only a year earlier. But the telegram reminded me that I should write some sort of report, for whom I didn't know, about the corals we'd found at the Solitary Islands. This was something I'd always intended to do but kept putting off until I had learned something of the subject.[2] I went to the university library to find a book that might help me put names to the corals. There weren't any, but there was a book, just published, written by Isobel Bennett, my childhood mentor, called *The Great Barrier Reef*. Maybe it would help? It did, but not much. I absentmindedly turned the pages. It was full of photos of a spectacular part of the world I had just said I wasn't going to work in – challenging work that would be all mine, in a place so important and so beautiful.

I was at a crossroad, an important one. Should I continue a career path in Canada on a subject I found enormously interesting and rewarding, and knew a lot about? I would be in the company of high achievers, part of a world-class team, and maybe even have a mentor. It was a once-in-a-lifetime opportunity, but would I have my much needed autonomy? Or should I start afresh, working on corals? I knew next to nothing about them, but it seemed that no one else knew much either.

I borrowed the book and hurried home. After much soul-searching with Kirsty I wrote two strange letters, both offering apologies for changing my mind. One to James Cook to say yes please and the other to Canada to say no thank you.

In the end, this was not an academic decision. Deep within me I'd always had a vague feeling of destiny, that I would some day return to marine life. Every few days there was something on television or in newspapers about the threats to the Great Barrier Reef from mining and oil drilling, threats to the most fabulous place I had ever seen. This was something I felt deeply about, something more important than just landing a job. My inner self tends to get the upper hand at such times and this was one of them.

In writing those letters I sowed the seeds for two records. The first was becoming the Great Barrier Reef's first full-time research scientist. The second was becoming a marine biologist without ever attending a single lecture on marine biology.

News of this decision turned Professor O'Farrell into a blubbering barrel of fury: if I was going to chuck in a career for a diving holiday, as he put it, then I could go to hell. He ordered me out of his office and told me never to return.

Had I done the right thing? With O'Farrell's words ringing in my ears I felt very insecure about leaving a project that had been so successful and enjoyable in favour of working on something I knew little about. Moreover, my little family was happy in a house that we'd turned into a comfortable home, one which was only a couple of hours' drive from Stonehenge. And although

Armidale was a small city, Kirsty could pursue her interests in music and acting there. Why leave?

My mother was horrified at the idea of us going to Townsville; it was so very far away and Noni had become the centre of her life. But move we did, in November 1972.

The Great Barrier Reef

Matters of politics

Some five years before we moved, at a time when I was busy collecting snakes, the attempt to open the Great Barrier Reef up to mining and oil drilling was raising its ugly head. It all started in 1967 when an enterprising sugarcane farmer applied to the local mining warden for permission to mine Ellison Reef, near Dunk Island, for limestone to fertilise his crops. He claimed the reef was 'dead'.

Fortunately for the future of the entire Great Barrier Reef, this caught the eye of the artist John Büstt, who immediately saw the ramifications of such a precedent and filed an objection to the farmer's claim. On what grounds could Büstt object? Political manoeuvres commenced that quickly developed into one of the biggest conservation battles in Australia's history. Both the Australian and Queensland governments claimed jurisdiction over the Great Barrier Reef but it transpired that, as The Reef is submerged at high tide and therefore not land, most of it is arguably in international waters, beyond the 12-mile territorial limit of sovereign countries.

It wasn't long before much of the Queensland coast had been quietly leased by the Queensland government for prospecting, leaving little doubt about the fate of The Reef should it fall into the hands of environmental vandals like Joh Bjelke-Petersen. And indeed, when Bjelke-Petersen was elected premier in 1968 he quickly became a conservationist's worst nightmare.

The mining warden required proof that Ellison Reef was *not* dead, and so began a string of conflicts that became both complicated and dirty. One after another, scientists, not wanting to get involved in anything that might tarnish their professional status, washed their hands of it, and even the longstanding Great Barrier Reef Committee, chaired at that time by Professor Bob Endean of the University of Queensland, declined to help. The University of Queensland even announced its 'official position': Ellison Reef was indeed dead. Where were the experts?[3]

It fell to Len Webb, a prominent rainforest ecologist; Eddie Hegerl of the newly formed Queensland Littoral Society (now the Australian Marine Conservation Society); Don McMichael, the newly appointed director of the Australian Conservation Foundation; and Vincent Serventy, the author of popular natural history books, to carry on the battle for the conservationists. In the end their efforts were unnecessary because the Queensland Minister for Mines stepped in and rejected the cane grower's application on legal grounds. There were celebrations all round, but they were short-lived because by 1969 the Bjelke-Petersen government had approved petroleum exploration licences over almost the entire Great Barrier Reef and had called for tenders to start drilling for oil.

This woke Australians up: it was a time I remember well, even though I was by then just starting my work on dragonflies. Most geologists supported the drilling, most biologists opposed it, and all politicians feared it because of the growing groundswell of public judgement. Ill-informed views raged back and forth: wasn't the Great Barrier Reef mostly just rock? Wasn't the best way of conserving it to use it for commerce? Wasn't oil actually good for corals? And so on. Opinion polls and newspaper editorials proliferated across the country and they made it clear that public opinion was decidedly against drilling.

As drilling rigs headed for Australia an all-powerful ally of the conservationists waded in. The Australian Council of Trade Unions, probably to goad the Queensland premier, banned the oil rigs and anything to do with them from all Australian ports. Many cheered, others thought it dangerous that unions had such power over governments. One outcome was an agreement to hold an inquiry into drilling, to be chaired by a judge and staffed by a petroleum engineer and a marine biologist. There were plenty of judges and petroleum engineers but where could an impartial marine biologist be found? The whole thing was shaping up to be 'the trial of the century', as one newspaper put it, until right at a critical point, in March 1970, the 58 000-ton oil tanker *Oceanic Grandeur* struck a rock in the Torres Strait, spilling its oil and creating a slick that started moving towards the far northern Great Barrier Reef. From then on, every oil spill on the planet got headline newspaper and television coverage and the inquiry was turned into a royal commission.

I was acutely aware of these events but slow to play any part in them, being preoccupied with so much else when my family and I first came to Townsville. Australia's 'long starvation of marine research', as Judith Wright put it, meant there were few scientists with qualifications in marine biology.[4] I certainly knew that, but had no idea what was in store when I was summoned to a meeting of the royal commission early in 1973. It was at the fledgling Australian Institute of Marine Science (AIMS), and had to be held outside, under trees by the beach, as there was no room big enough in any of its buildings.

Someone from Canberra, sweating profusely in his suit and tie, chimed up, 'Ah, er, Dr Veron – doctor, is it?' I nodded. 'I'm surprised we haven't heard about you before – what exactly is your field of expertise?'

I explained my PhD on dragonflies.

'So you have no qualifications in marine science?'

'No.'

'Then anything you say will be inadmissible. Thank you for your time.'

That was a slap in the face that didn't go away.

Although he continued to cling to power, the conservation victories of 1974 substantially eroded the authority of Joh Bjelke-Petersen. As a last resort, the Queensland government issued a High Court writ restraining the Commonwealth from making laws affecting Queensland's coastline. However, by mid-1975 it was mostly over: the then Prime Minister, Gough Whitlam, pre-empted the writ by declaring federal ownership of the Great Barrier Reef, henceforth to be a marine park, and the High Court promptly backed him up.

Well do I remember that time. I might have been useless to the likes of a royal commission and I had no idea what Whitlam was contemplating, but for months I'd been getting strange phone calls from Canberra parliamentarians about corals, and more particularly, where reefs occurred.

'So, eh, Dr Veron, let me get this straight. You say corals can be found where there are no reefs, right? And that the . . . uh . . . Great Barrier Reef actually extends to the coastline, even if there aren't any reefs there, right?'

And so on. I knew something was afoot, but Whitlam's proclamation took me by surprise, as it apparently did everybody else. Although I had helped, all this made me feel ill at ease. So little was known about coral, or in fact anything to do with The Reef. There was nobody to turn to, and almost nothing had been written about most of the questions that were put to me, not to mention the many more I was asking myself.

The tropics

When Kirsty, Noni and I first arrived in Townsville it had a population of only 72 000, although that made it the second-largest city in Queensland, a state almost twice the size of France and Germany combined. I imagined the university would be beside the sea with a nice view out to the Great Barrier Reef, but it was well inland, amid flat scrubland covered with chinee apples – nasty thornbushes. The city itself was nothing like it is now. There were hardly any trees in the suburbs, Cyclone Althea having devastated the place the year before our arrival, and most houses still looked barren and uncared for. And Kirsty soon discovered that as far as music and the stage were concerned she had none of the connections she so much valued in Armidale. However, we soon found a good side – people never needed to take the key out of their car, or lock their house, and almost everybody wore thongs, shorts and T-shirts everywhere, which suited us fine. And it was a very friendly town.

We'd been advised to avoid coming in the wet season, when the Bruce Highway, the only access road to Townsville from the south and in places just a single lane, was frequently closed for weeks on end. But as my post-doc could be started any time we decided to risk it in November, mostly because I couldn't wait to begin my new job and O'Farrell had all but thrown me out of my old one anyway. We arrived without mishap, and as expected we found the city oppressively hot after the cool climes of Armidale. We were installed temporarily in a house with so many cane toads around it that the ground seemed to move at night.

The day after our arrival I was wandering around the university's biological sciences department when I spied a door with the name Professor Cyril Burdon-Jones on it. He was the head of the department and the man who'd written to offer me my job. After getting his secretary's uninterested nod I knocked and went in, dressed in my newly acquired Townsville regalia. The professor, a rather supercilious Welshman who put great store in formalities, didn't seem to appreciate my tropical attire. Almost everything about him contrasted with everyone else in his department, half of whom were poms he'd recruited on coming to Australia. And so he kept his distance even from his recruits, who were a laidback and friendly lot and generally no better dressed than the locals.

'G'day Prof, I'm Charlie Veron.'

'I see,' Burdon-Jones replied in a disdainful tone, an eyebrow slightly raised. He gave me a cool handshake, waved vaguely in the direction of the Great Barrier Reef and announced, 'Your job is to go out there . . . and do something.' As an afterthought he added, 'And try to stay out of trouble.'

That was in fact the only job description I ever received, which meant I had a free hand – just what I wanted – but then neither of us knew anything about coral or the Great Barrier Reef. As it turned out, that day was the high point of my relationship with my boss for a long time. Over the ensuing nine months Burdon-Jones repeatedly announced, for no reason I could see, that he would be finding someone 'more appropriate', or 'more suitable' for my job. Not exactly what I'd hoped for, but at that time I had more important things on my mind.

After six weeks of living with the cane toads, which Kirsty found unnerving, we bought our own house. It was a roomy new fibro box on stilts, in a suburb of roomy new fibro boxes on stilts, just like the song about ticky-tacky boxes except there was no hillside, only flat clay. Certainly an aquarium was called for, but that was my thing; what was really needed was something for Kirsty. We had no money, so we bought a piano. Kirsty's instrument was the violin, but she'd also learned piano at school.

New friendships came easily at James Cook University, and several people Kirsty and I met there became friends for life. One of these was Alastair Birtles, a large, bearded Englishman who had graduated from Oxford and then made some fascinating journeys to exotic places like the Red Sea. When we first asked him to dinner we discovered that he never did anything on time. No matter, in his normal style, which we were soon to get used to, he arrived only a day late. He was also late in leaving, having bunked down in our spare room, which became his room for the next nine months. Al was a welcome addition to our family, always interesting and always ready to pitch in when something needed doing.

Kirsty and I were still trying to cope with Ruari's death and had decided the best way to do that was have another baby. Kirsty was already pregnant by the time we left Armidale, but soon after we arrived in Townsville she started to get repeated bouts of bleeding. The medical advice was to take it easy, but that changed when she began having contractions: she was sent to bed to rest. When the contractions got worse her doctor prescribed four ounces of Bacardi rum every two hours. At first one or two doses did it, but soon the treatment had to be kept up for longer, sometimes for days on end.

Being confined to bed in the oppressive heat and humidity of a Townsville summer, and at the start of a new life where there were no old friends and no family on hand for support, had a devastating effect on Kirsty. Gradually she withdrew from the day-to-day world and became less and less interested in well-wishers, until even reading to Noni became a chore. As the weary weeks unfolded she subsided into a state of confusion and disorientation. When she finally lost the baby, at the beginning of March, she was a mental and physical wreck with few feelings and no interests.

The effects of this prolonged struggle on top of Ruari's death turned Kirsty's life into an ordeal from which she took a long time to recover, and when she did we decided enough was enough – we would not have any more children. The only good side of the ordeal was that both Noni's grandmothers came to help. We were able to get Noni into kindergarten three mornings a week, but most of the rest of the time she had one or other grandmother to play with. My mother was reluctant to return to Sydney at all until Kirsty was completely back on her feet, but also because she couldn't bear to part with Noni, who was developing into an extraordinarily colourful character. It was a mutual love. Noni adored her gran, as did Kirsty – all part of the wonderful magic of my mother.

Throughout this turmoil I had to at least try to learn something about corals. The university library, as it turned out, had foreseen the potential for reef studies and purchased many books about marine life, including some about corals. These weren't guide books, which didn't exist in those days, but ancient monographs about taxonomy, mostly long and complicated and full of unexplained jargon. I despaired of most of them – they would have been no less incomprehensible to me had they been written in Latin, as indeed several were. So I decided I would try to do with corals what botanists often did with plants: map the community types, work out what the dominant species were, and classify communities accordingly.

This was not as straightforward as it sounds, because it depended on my being able to identify the dominant species, a task which inevitably turned me back to the daunting old monographs. But gradually a sense of determination took over. I would go out on the university's new research vessel, the *James Kirby*, to local reefs to see the corals in real life, collect them, and then return to see what the monographs had to say about them. I had plenty of helpers on these trips as by then scuba diving on The Reef had started to become popular with students and they all wanted to come with me. This soon created a problem for the university: wasn't diving dangerous? I kept being asked about regulations and qualifications. All such matters have a solution,

then: I agreed to be the university's diving officer
[for] anyone who'd taken a diving course and knew
[...] doing permission to go diving. I believed then,
[...] for divers with sufficient experience, safety is a
[...] responsibility, not regulation.

This solution made everybody happy. A lot of students accompanied me on many trips and before long I had thousands of corals to play with. That was all very well, but having a lot of corals wasn't much use without knowing what they were, something I was clearly unable to work out. I could soon recognise dozens of kinds of coral and give them nicknames, but when I tried to identify them from the taxonomic volumes that meant something to me, I always came up with a litany of vague alternatives.

As I quickly discovered, the essential problem was that corals on reef slopes vary their growth form and skeletal structure according to the environment in which they grow. Thus a familiar coral growing on a wave-hammered reef front might be compact with short stumpy branches. If that coral is common it's a simple matter to follow colonies of it down the reef slope to deep water, whereupon the same species gradually turns into something quite different, perhaps developing long branches. These different forms had been given different names by the experts because, I realised, without scuba these experts had never seen corals as I was seeing them. To make matters more complicated, big colonies regularly have one growth form at their top, different growth forms on their sides, and different forms again at their base. These growth forms had often been given different species names.

I decided to work these variations out – I had to, so that I could separate a single species from a group of related species – but I gave up trying to stick names on them. I lost faith in the idea that any clear picture of Great Barrier Reef corals would emerge if I relied on historical taxonomy: I would have to start from scratch, observing what could be seen on reefs rather than what was written in monographs. Eventually I decided to stop calling my work taxonomy altogether and instead call it population ecology. That at least seemed an escape route.

Had I made the wrong decision in abandoning insects? I began to seriously think so.

As luck would have it, the Second International Coral Reef Symposium was to take place in June 1973, and it was to be an event like no other. The organisers, the Great Barrier Reef Committee, chartered a 10 000-ton cruise ship, the *Marco Polo*, to take participants from Brisbane to Lizard Island in the northern Great Barrier Reef and back to Brisbane.[5] This amounted to ten days of talking day and night, drinking beer, diving, and generally having a good time.

There were 264 scientists aboard: a perfect opportunity to meet the most important people in the business, especially for me, a novice. The first important person I met was David Stoddart, a freckled, red-headed, somewhat stout character with a deep resonating voice and pure Cambridge accent. By that time he was a seasoned and well-known reef geographer.

'My dear fellow,' said David, 'how very nice it is to meet you. Sorry to have to tell you, but your paper was rejected – Pat Mather's doing of course. *I* thought it the best of the bunch. But she doesn't approve of people like you messing around with taxonomy.'

'Oh?' Pat Mather was a curator at the Queensland Museum.

'In fact yours was the *only* paper to be rejected,' he said, 'but never mind, I've put it in anyway. Pat won't notice, damn silly woman. More to the point,' he continued, 'why don't you join our expedition later this year to the northern Great Barrier Reef? Bring the *Kirby* – go right to the top end. Would be great fun, eh? Don't forget to bring lots of booze, there's a good chap.' Then off he went.

A few days later, I was sitting at a table on the aft deck having breakfast and nursing a hangover when a middle-aged lady with large glasses walked up to me, smiling.

'Charlie Veron! What are you doing here?'

I had no idea who she was. I could only sit and stare at her.

'I'm Isobel Bennett.' She watched, still smiling, as that slowly sank in.

'I'm working on corals,' I said, for want of anything else to say.

'Really? I thought you'd come back to the sea. You went to Armidale, didn't you? I lost track of you after that. Well, catch you later.' She disappeared.

As it turned out we didn't have that catch-up; in fact I didn't see her again that trip. Need I say that she left me curious as to how she'd recognised me. I was twenty-eight by then and had no distinguishing characteristics, except perhaps a husky voice and bushy beard. The only other time she'd seen me – when my mother took me to visit her – I was eight. I wondered about Issie, as everyone called her.

The *Marco Polo* continued on to Lizard Island, stopping at places of interest where it was met by launches to take divers out to pre-selected spots. The weather was mostly bad, the currents were mostly strong, and the divers mostly beginners. There were many rescues, one involving about ten divers strung out over hundreds of metres of open ocean, caught in a current and waving frantically as sharks (or so they no doubt imagined) circled. Sharks or no sharks, it's amazing that everybody lived through that symposium.

On the last night, after we'd left the southernmost reefs and were heading back to Brisbane, the whole ship's company met on the aft deck for a farewell buffet dinner and 'ball'. That company included Cyril Burdon-Jones. I was on the dance floor when he walked up to me.

'Enjoying yourself, are you? I didn't think much of your paper, so I'm thinking I should find a more suitable person for your position.'

That sure took me by surprise. My paper had been well received, but that aside, I felt that his threat to sack me yet again, at that particular moment, was simply evil.

If I'm going to get sacked I may as well do it in style.

I grabbed the professor by his tie and shouted loudly that I was going to throw him overboard. It must have looked rather comic – the normally dignified Burdon-Jones being dragged across the dance floor, struggling like a spaniel on a leash. Some revellers around us started clapping; the professor was finding out what young people thought of him. At that point I let him go and he scurried off, not to be seen by anybody for what remained of the trip.

Once back in Townsville I wasted no time seeing Ken Back, the vice-chancellor, to get his approval – on the off-chance that I would still be on the university payroll – to use the *James Kirby* for the last phase of David Stoddart's expedition. This wasn't my first meeting with Ken. A few months earlier he'd summoned me to his office to haul me over the coals for writing to the local newspaper about a nickel refinery's plan to dump wastewater into the ocean just north of Townsville.[6] His beef was that I'd used the name of the university in my letter and shouldn't have. I saw his point, but left his office talking about Kirsty and how my post-doc was going. I had the feeling that I'd met a friend.

Ken and I talked a long time about Stoddart's proposal. He said he would discuss it with Davie Duncan, the skipper of the *Kirby*. The main part of the expedition was to take place in the central Great Barrier Reef; our part involved the *Kirby*, with me in charge, collecting Stoddart and some others and heading north to the Torres Strait, exploring the outer reefs as we went. It would be the first research expedition to the northern Great Barrier Reef.

Underwater utopia

The Great Barrier Reef is known throughout the world as one of the greatest natural ecosystems on our planet. It is the largest and most spectacular structure built by living organisms on Earth, its size virtually impossible to comprehend. The first astronauts saw The Reef from outer space, the only living thing on Earth they could see. I've not had that pleasure, but I have seen it many times from the air. One of those times, thirty years ago, I had a bird's-eye view from the nose cone of a Neptune bomber, sitting beside the then minister for science. As the plane flew on, hour after hour, one reef region after another came and went in what, for the minister, must have seemed like an endless progression of more and more of the same: he kept falling asleep. Poor man, he was frequently woken by my outbursts of enthusiasm as one favourite dive site after another came into view, yet when I retraced our flight path on a map after landing, we were both reminded that we'd covered but a fraction of the whole. Even from the air the Great Barrier Reef is too big to take in.

Diving on The Reef is something else again. Only the best movie photographers can capture the ambience of pristine coral gardens. These contain a profusion of life: corals, soft corals, fish, anemones, urchins, starfish, shellfish, and little creatures everywhere, a diversity never seen on land, not even in rainforests. Then there are the underwater cliff faces and ceilings of caverns ablaze with the colours of filter feeders: ascidians, sponges, sea-fans, crinoids and more soft corals, most beyond the knowledge of science.

Reefs in poor condition feel like graveyards, because divers hear only the bubbles from their own breathing. Between breaths such reefs are morbidly silent. Healthy reefs are the opposite; the healthier they are, the noisier they are, because they're full of little animals all with something to say. This is especially so at night, when even a torch beam can start a cacophony of crackles, chirps and croaks. Some come from fish and are identifiable, but most come from unseen little critters, of which there are thousands of different kinds.

To dive at night is to enter a spectacular world, for the diver's torch reveals these tiny animals by the thousand: iridescent clouds of plankton of all shapes and sizes, swimming frantically, lured by the light. At night corals open their tentacles to catch these little animals, transforming themselves into anemone-like creatures with long sinuous tentacles, quite unlike their daytime guises. At night, too, large predators of every description move through the distant darkness. Sharks roam unceasingly, sensing the presence of anything moving in the dark.

Photographers sometimes capture these scenes, by day and by night, but they can never convey the emotions of those of us fortunate enough to have looked upon this world first-hand. The greatest reefs of the Great Barrier Reef always instil awe and wonder. The sight of large animals – whale sharks, manta rays, giant groupers, a dozen kinds of big, ocean-going silver fish, sharks and the occasional whale – never fails to thrill. But there's also fear, especially on the lesser-known outer reefs of the far north. The threat of depth is ever-present, for clear-water reefs that plunge down into deep ocean can be deceptive, a fatal attraction for a tourist wanting to get just a little more out of a trip of a lifetime. Sadly, a spectacular reef face is the last thing some ever see.

Electronic navigation and satellite imagery have virtually put an end to The Reef's reputation for being one of the world's most dangerous places for ships. Today, tropical cyclones are more feared than reefs, but there was a time when the Great Barrier Reef had more shipwrecks than the rest of Australia's coast combined. A glance at any chart shows why, for there are thousands upon thousands of reefs of all shapes and sizes hidden just below the surface and enveloped by strong tidal currents. These form complex mazes from which, in a storm, there is little chance of escape without modern navigation and powerful engines. I have come across dozens of ghostly wrecks on The Reef, usually unexpectedly and always making me wonder what happened to the crew.

Although I've worked in almost all the major reef regions of the world, most of the exceptional dives of my life have been somewhere on the Great Barrier Reef.

My studies of corals started taking shape, and the more I saw of the Great Barrier Reef, the more big-picture questions arose. How did reefs grow? Why did they exist at all? Why did corals do all this building? Did they need reefs? Why did corals make colonies with such spectacular architecture? Why were so many different kinds of just about every major marine animal group to be found on them? Had it always been so?

We have answers to most of these questions now, but when I started thinking about them it seemed like trying to put together a dozen great big jigsaw puzzles all at once, with the majority of the pieces missing.

'The Great Barrier Reef is only ten thousand years old,' a colleague of mine announced during a seminar thirty years ago. She was both right and wrong, because there's another player in this game: sea level change. Sea level change makes the Great Barrier Reef both old and young irrespective of geology or biology. It depends on one's point of view.

John Chappell, a reef geomorphologist at the Australian National University, lateral thinker and good friend, had been making intriguing discoveries about sea level changes for decades, not on the Great Barrier Reef, but in Papua New Guinea, for when it comes to sea level you can work wherever uplifted reefs are best preserved, as the sea is almost the same level everywhere. During the ice ages the sea level went up and down in response to massive polar ice shelves forming, melting and re-forming, taking thousands of cubic kilometres of water out of the ocean and then returning it. In resolving the complexities of sea level changes, John took his inspiration as much from coral ecology – the living veneer seen underwater – as from geology. Certainly, it is the life on a reef, overwhelmingly corals, that builds reefs, but it is mostly the power of waves that cuts them down. Both processes are slow, taking hundreds of thousands of years, far slower than sea level changes, so reefs during the ice ages continually played catch-up with the sea. Some lost the race and drowned because the ocean was too deep when they started growing and their essential life support, light, was cut off.

Glacial cycles have changed the sea level, but only by about 150 metres or so and only during the Ice Age. Far greater changes result from sea floor spreading as continents move. That is decidedly geological and is full of crises. At one point in time the Mediterranean was not a sea, but a gigantic hole some 1500 metres deep on average and much deeper in some places. This hole was filled when the Atlantic broke through the Strait of Gibraltar 5.3 million years ago, forming the mother of all rapids, a single event in geological time that must have caused the sea level to drop enough to kill every coral that grew on a reef flat the world over.

Little did I know in my early years as a scientist that changes in the Earth's orbit around the sun – Milankovitch cycles, which pace glacial cycles in much the same way as a pendulum paces the speed of a clock – are another ingredient in this already heady array. Milankovitch cycles control glacial cycles, which control sea level changes, which make reef corals ephemeral, super-competitive, and addicted to algae: the story of coral reefs does indeed go all the way from astronomy to cell physiology in a single string of processes varying in time intervals from millennia to days.

I had something else to puzzle over in my early years, closer to home. Were reefs biological or geological structures, and if they were both, where to draw the line? In 1976, two professors at the University of Queensland edited a two-volume book about coral reefs, one volume devoted to biology, the other to geology. These emphasised the division, and the problem became worse when I read an article in which the author claimed that reefs were 'fragile'. Reefs fragile? That was like calling a dam wall fragile. Argument about this separation of biology and geology went on for at least two decades, generating continual controversy.

We now understand that the living veneer of a reef is indeed fragile, because it is restricted to the interface of land, sea and air, zones where nothing else survives because they are the most violent places on earth when storms arrive. It is this living veneer that gets consolidated, by yet other types of algae and by the chemical action of rain and seawater, into a solid limestone reef, which is anything but fragile. That fragile organisms should build such enduring structures is one of Nature's great paradoxes. We need to take a close look (as we later shall) at the ecological environment of corals to see how this is done.

I never researched any of these subjects personally, I just thought about them, mostly by sitting in a dingy with my toes in the water or on a beach, gazing out over a reef: a great way to occupy a lot of time.

Reef builders

Coral reefs the world over are mostly built by coral, something they've been doing for hundreds of millions of years. This makes them the greatest builders that have ever existed, only to be pipped at the post very recently by us. Despite their extraordinary achievements, a coral is a simple organism, with the basic design of a sea anemone. Some have soft bodies but these are only very distantly related to hard corals, the Scleractinia, and there are many other differences, the most important being that it is the hard corals that build reefs, using the energy-capturing ability of green plants to do so. Hard corals, unlike anemones and soft corals, have skeletons of limestone, all with intricate designs. Some live alone and so are called solitary corals, but most that are found on reefs form colonies consisting of hundreds to thousands of tiny individual animals called polyps, all working together to produce the architectural masterworks we can see.

It has long been known that reef corals (sometimes called hermatypic or zooxanthellate corals) have minute algal cells (zooxanthellae) inside the cells of their inner body layer, and that these tiny algae produce nutrients by photosynthesis. Len Muscatine, an inspirational American coral physiologist, was making great strides in this field when I was doing my early work on corals, claiming that algae produce most of the nutrients that corals need.

At the time, I found this hard to believe because most corals are exceptionally good at catching their own food. During the day, polyps usually hide inside their limestone skeletons, but at night they show themselves, body and tentacles, looking, as I have said, just like little anemones. They don't do this for fun, they do it to capture food, and that's something easily seen. Corals in fact have voracious appetites, ensnaring tiny zooplankton in their tentacles, which are often armed with batteries of stinging cells. They stuff plankton into their mouths one after another, just like the anemones at Long Reef did with the worms I fed them.

But Muscatine was right – a habit of his. Experiments have shown that most nutrients corals use do come from their algae. It's an intimate symbiosis. At least, this is the case with zooxanthellate corals, but there's another group of hard corals, the azooxanthellate species, which have no algae, and these are mostly found in caves, where they only need reefs to supply zooplankton to eat. Azooxanthellate corals also occur in very deep, near-freezing water where there is no light at all. These groups are usually easy to tell apart: most zooxanthellate corals form colonies whereas most azooxanthellate corals live as tube-shaped solitary individuals. The symbiosis between coral and their algae is of enormous importance: it is the key to their capacity to build reefs, and as we will see, when that symbiosis breaks down it can cause mass death.

Reef corals, like forest trees, live in highly competitive ecosystems where corals vie for sunlight, often by forming sunlight-capturing plates, architectural masterpieces that make the most of the building material available and permit rapid growth. Branching staghorn corals do likewise, sometimes growing as much as 30 centimetres a year. These corals can be readily broken by storm waves or eaten by predators, so other species have a different strategy, forming solid structures that waves seldom break and predators seldom eat. While they're small, these 'massive' corals, as they're called, might be outgrown by the plates and staghorns, but they bide their time, waiting until storms and predators clear away the competition. Eventually they grow large enough to avoid being overgrown and then they get a continuous share of sunlight.

All this can readily be observed by divers, but what happens if a zooxanthellate coral is deprived of light? One of the first observations I made when I first dived on reefs was that the poorer the light (whether from turbidity, shading or depth), the slower the corals grow, until they don't grow at all. Consequently the number of species dropped off the deeper I dived. Green plants do much the same thing where sunlight becomes scarce, but most green plants can't capture live food whereas corals can. So what's going on? Why are corals so totally addicted to their algae? I still ponder that question. The bottom line is that if any zooxanthellate coral is deprived of light they eventually die, no matter how much food they have. Corals are not on their own here; giant clams also capture food by filtering seawater, and like corals can grow on nothing other than seawater and sunlight because they also have photosynthetic symbiotic algae, albeit of a very different sort.

Light is not the only thing corals compete for; they vigorously compete with each other for space. At first glance they appear to grow harmoniously together, but on closer inspection, especially at night, they can be seen attacking each other, using their tentacles or, more commonly, bundles of long filaments extruded through their mouths. As soon as a filament touches another coral the battle is on, for each tries to sting or digest the tissues of the other. Peaceful-looking coral gardens are anything but peaceful; there are war zones everywhere.

But corals aren't always competing. When it comes to reproduction they go to extreme lengths to co-operate. Across the entire Great Barrier Reef, one or two nights each year are like no other, for then mass spawning takes place. Like most plants and animals, corals need to be able to reproduce sexually, but as they can't mate like mobile animals, they release eggs and sperm, usually in bundles, which float to the surface where cross-fertilisation can take place. This must be synchronised to within just a couple of hours, something governed by the time of sunset, the phase of the moon, and changes in water temperature. When every polyp of every colony of a whole coral garden spawns in synchrony it's show time, for egg and sperm bundles are released in such quantity that the water column looks like a pink snowstorm, except that everything goes up, not down. Fish in frenzies eat their fill but still the show goes on, and not just with corals; other organisms, picking up on chemical cues, cash in on the timing of the coral spawning when fish are full of food and release their gametes also. The ocean surface becomes covered in slicks of pink spawn, all to be swept away by surface currents to another reef, or to oblivion. Mass spawning is a sight never to be forgotten, at least not for a year, when the whole show is played out once again.

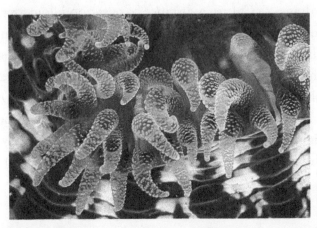

Batteries of stinging cells on the tentacles of a coral.

Expedition north

At the time of the Stoddart expedition (1973), it's fair to say, nobody had advanced any clear idea about why the Great Barrier Reef is where it is or how long it has been there. But by the mid-1980s, about a hundred articles on its origin had been published, the consensus being that it was of recent age, mostly post-Pleistocene, meaning post the last ice age, as any pre-existing reef would have been removed by erosion. This idea followed work done in the Caribbean by the American geomorphologist Ed Purdy, who in 1974 hypothesised that reefs are composed of successive layers of limestone, each layer growing in response to a sea level rise.[7] Not only did this create the internal morphology of reefs, as seen in seismic profiles, but it also controlled their modern shape and position.

As soon as Ken Back gave us the go-ahead for the expedition, a small group of us immersed ourselves in planning. We needed to go in December, when the monsoons came and the seas were usually calm, despite it being cyclone season. Calm seas or not, our plans were a bold move, for the charts of the time were frighteningly archaic: from *Capt'n Matthew Flinders as amended* (1802).

The places most of us wanted to visit were the ribbon reefs, the outermost reefs of the northern Great Barrier Reef, which form a spectacular chain starting east of the Torres Strait and running south almost to Cairns, a distance of over 700 kilometres. Tijou Reef looked especially interesting because, judging from depth soundings of the time, the abyssal depths of the Queensland Trough come closest to The Reef at that place. Just how far from the ribbon reefs was the Trough? This question was very Darwinian in scope and thought, for it begged a connection between the origin of the Great Barrier Reef and Darwin's theory of the origin of atolls.

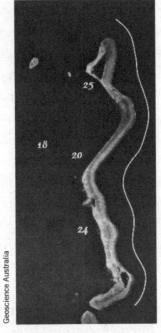

Tijou Reef. This long ribbon reef, part of the outer Barrier, is about 1000 metres wide. The white line to the right shows the 1000-metre depth contour; the numbers to the left show depths in metres.

On the first leg of our expedition, we stopped at several islands where divers could work on corals and others could sample beach rock, reef flat corals and clams suitable for isotope ageing. We then headed up to Lizard Island, where the building of the now famous research station had just commenced, then on into almost unknown territory. After anchoring the *Kirby* on the western side of Tijou Reef, we crossed to the outer face in a tinny for a dive, and so became the first people, as far as I know, to dive anywhere on the outer face of the entire northern Great Barrier Reef. With the water crystal-clear we could see the reef face, spectacular in its grandeur, plunging steeply down and down. It was like diving down a mighty dam wall and it soon became obvious that this outer face of The Reef *was* the western edge of the Queensland Trough: the two were one. This did not sit well with any of the hypotheses later put forward about the age or origin of The Reef. To have maintained themselves in such a position, the ribbon reefs had to be old, very old, even as old as the Trough itself. It was during that first dive that I knew the northern Great Barrier Reef had a history nobody had even vaguely contemplated.[8]

That outer face is a deathtrap for divers, at least it was until dive computers, which sound warnings, came into use. On that first dive, I was so absorbed in the splendour of the reef face that by the time I checked my depth gauge I discovered I'd reached 50 metres, nearing the maximum safe depth for scuba, and I was going down rapidly. This caught me unawares because even at that depth the coral communities looked like those found on wave-hammered upper reef slopes. I've never been on the outer face of a ribbon reef when the sea is very rough, but judging from the corals the turbulence must be extreme. It's only during the monsoon season, at the height of summer, that the sea goes calm, and then it may go so calm that it's completely flat, without even a ripple. At that point the horizon disappears and the water surface reflects the sky, hiding everything beneath. If this happens at high tide the reefs are at their most dangerous, for boats get glassed in; it's as if they're floating on a gigantic mirror. Many times we drove a zodiac in front of the *Kirby* to warn Davie of a reef ahead.

At low tide, when the outer reef flat is high and dry, the ribbon reefs look very unlike most other reefs, for they're so wave-pounded that there are no corals, just hard, consolidated, flat limestone pavement. So hard and flat that I imagined it might be possible to land a jet on it with impunity.

Diving on the outer face of Tijou Reef was memorable for another, absolutely extraordinary reason. A couple of us were down about 20 metres or so when we were hit – there's no other word for it – by a blast of the deepest and most intense sound imaginable, as if we were in front of the biggest pipe of a giant organ. The same thing happened again the following year. I had no idea what it could possibly be until, many years later, I talked to a researcher who worked on whale sounds. He said it would have been a whale, probably a sperm whale, checking us out, doubtless because they hadn't come across divers before. I'd never heard of this happening to anyone else, and nor had he. Maybe most whales today know about scuba divers and ignore them? The answer still eludes me.

We left Tijou Reef in mid-December and headed north to the remotest island of the Great Barrier Reef. Perhaps some historians would have known about Raine Island, but all we could find out about it was that it had a stone beacon built by convicts in 1844 to warn sailing ships away.

When we arrived we felt we'd rediscovered the island: 32 hectares of low sandy cay with the most incredible birdlife any of us had ever seen, as well as the biggest green turtle rookery on Earth, as far as we knew, with at least 280 turtles heaving their massive bodies up the beach at dusk. (Our discovery was dwarfed the following year at the same location when we counted nearly twelve thousand turtles in a single night, a world record for green turtles.) We watched these beautiful animals all night, for there is something compellingly emotional about a turtle nesting. Everything these animals do requires enormous effort, and they constantly seem exhausted as they search for a good nesting spot before they start digging, then laying over a hundred eggs in a pocket scooped out by their hind flippers. Finally, these nests must be covered up before the mother slowly makes her way back to the ocean and swims away.

Underwater, we saw turtles stacked three-deep on reef ledges, waiting for the night to come, and from the *Kirby* we saw several thrown high out of the water as tiger sharks ripped their fins off, the exhausted females being easy prey.

Our accounts of this place on our return attracted a lot of attention and led to the start of the Raine Island Corporation, resulting in the island being given the strictest conservation protection in all Australia. I returned to Raine Island several times, having been given special permission to go whenever I chose, a privilege that ended when the work of the corporation was turned over to the Queensland government in 2005. No grumbles there, I'm just happy to see the place protected, since anybody who goes ashore sends up clouds of birds and disturbs nesting everywhere.

Sadly, I believe this spectacular place and the life it supports is doomed because it has little chance of surviving the sea level rise now under way. This view is unpopular with most of the scientists and managers engaged in a battle to save the turtle rookery by building up the cay with sand from the mainland. The island is surrounded by deep water; will the sand survive a cyclone? I think not, especially at higher sea levels, but I suppose anything that might buy precious time deserves a try.

Further north we discovered the most formidable and spectacular reefs I have ever seen. These are the deltaic reefs that form the northern limit of the ribbon reef chain, and which in aerial photos look like a string of wide river estuaries, opening east into the depths of the Coral Sea. To the west there are vast areas of deltas, complete with progressively smaller tributaries.[9] We had a set of World War II photos, the sum of all knowledge of them at that time. The mudflats forming these deltas, as we initially supposed them to be, turned out to be solid limestone with the tributaries, in the form of deltaic patterns, carved into them. The channels were all U-shaped in cross-section, with vertical sides and flat, rubble-covered floors.

What was most spectacular of all about these reefs were the currents on the ebb tide, for they were ferocious, the current in the channel we were in almost bringing the *Kirby* to a standstill, engine roaring at full throttle. Davie Duncan was a great skipper; we found an anchorage.

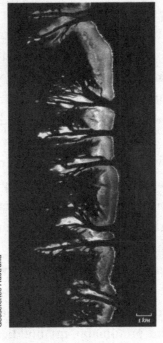

The deltaic reefs north of the ribbon reefs. The white parts look like mudflats but are solid limestone; the dark 'rivers' are channels cut by tidal currents.

'Bloody hell, Charlie, coming here was *dumb*,' he said as he wiped sweat from his brow.

Davie's got no sense of history. Nobody's ever been here before. Or if they have, they never lived to tell the tale.

Our first job was to confirm that these reefs were indeed solid rock and not mud; the second was to check out the channels. There was nothing for it but to start a scuba dive from one of the small back tributaries and let the current take us – three of us, with little marker buoys attached and followed by a zodiac – down the channels as they became bigger. We soon found ourselves going faster and faster with no chance of stopping. Although wildly turbulent, the water was very clear, giving no sense of movement, but in it were unseen forces of compelling strength, pulling us up then down, this way then that, sometimes spinning us full circle. The channel edges, almost devoid of coral, whizzed by and even the bottom, which we could glimpse deep below, appeared to be on the move. By the time we reached the main channel things were getting a little scary, but the sound of the zodiac's outboard above us was reassuring. I was thinking it was time to call it quits and surface when a huge ridge of reef loomed ahead, forming a sort of sill, which we later discovered was the outer lip of the channel, about half a kilometre wide. Over the sill we went, at terrific speed, and then down down down, losing sight of each other as we went. Even the bubbles from my regulator went straight down. I inflated my vest at about 30 metres and, relieved that this arrested my descent, swam for the surface, searching for my buddies as I went. It was a daunting place, especially as there were hundreds of sharks all moving very fast, as if it were a feeding ground.

The surface, when I reached it, was a lonely place. My buddies were nowhere to be seen: all I could see were big, foam-topped standing waves, holding their position against the current, which was pushing me, and presumably my buddies, relentlessly away from Australia. There was nothing for it but to wait, and hope the others had reached the surface. Fortunately, our boatman had his wits about him. Our marker buoys had disappeared when we were pulled down, so he zigzagged back and forth until he found each of us. I have seldom been so pleased to see a boat, with the others safely in it, in my life.

Even for a zodiac with a powerful outboard, the return trip over the waves and against the current was slow, and once back in the channel we had traversed it was fascinating to see whirlpools and smooth mounds of upwelling water forming here and there and then disappearing as we sped along.

Diving when you have control of what you're doing is one thing, but when you have none it is quite another. I know of nobody else who has dived in these channels. No surprises there, but at least we gained an impression of what they looked like. Later this gave me some ideas about how they had formed: from tidal currents, erosion, sea level changes and a subsiding continental shelf.[10]

Mer Island, known as Murray Island when we were there, lies to the north of the deltaic reefs. Legend had it that it was a dangerous place to visit as its occupants were in the habit of inviting unexpected arrivals for dinner. I had little doubt that they once did, especially after their chief, a massive man the colour of obsidian, bellowed with laughter as he told us how his father delighted in cooking and eating the tongues of shipwrecked sailors, or nearby Darnley Islanders when no sailors were around, while their owners were made to look on. Still chuckling, he invited us ashore. It was Christmas Eve and the island's children were going to stage a concert.

We made our way to the village hall, keeping in mind the chief's parting hope that we wouldn't be turned into a traditional feast. The hall was a large corrugated-iron shed and there we waited, and waited. We were about to give up when the distant sound of children singing came wafting through the coconut palms. I bolted for our zodiac, not for safety, but to get my tape recorder. I decided to sacrifice Beethoven in favour of the children's songs, but at first I had no idea how good that concert was going to be. The children filed in, dressed from head to foot in their best traditional Papua New Guinea-like regalia, and for hours – almost until dawn – they sang and danced, and I recorded the whole thing. I have never forgiven myself for throwing out, twenty years later, what I thought were a lot of old Beethoven tapes. I had in fact thrown out hours of children singing in Meriam Mir, a language that at one stage looked like it was on a path to extinction.

On our last evening at Mer I found myself on the main beach, sipping beer with a guy called Eddie Mabo. Eddie (1936–1992), now enshrined in Australian history for his spectacularly successful legal moves to give indigenous people ownership of their own land, was all grumbles. Certainly he complained about the white man's law that said his people did not own their own land, but he also complained about his own people. He wanted to build a tourist resort on a small island adjacent to Mer, a pretty place shaped like a horseshoe and which sometimes had bubbling water in the middle, the last murmur of an almost extinct volcano deep below. Eddie admitted the island was sacred and that the islanders wanted none of his ideas. Most of them, he added, wanted him banished from Mer altogether. That's a bit of the history I've never seen in print.

Unfortunately David Stoddart had to get back to England before we headed to the ribbon reefs, so we put him ashore at Lockhart River Mission, halfway up Cape York. Although he couldn't participate in any of our diving – he was no swimmer – he spent his time surveying vegetation on the little coral cays we visited and was generally a delight to have on the boat, mostly on account of his incredibly old-world sense of humour and his colourful language. His deep booming voice, which continually echoed around us, was an amazing mixture of upper-crust Cambridge and the pits of sailor-swearing, which, no matter how blue the words, always ended up sounding like poetry.

His last day's work done, a few of us went ashore to collect him. He looked suspicious as we approached, as well he might. We pounced on him, stripped him, rolled him in some mud, dragged him up the beach, rolled him in sand and sat him up. I then photographed him grinning at the camera with just his hat on.

The expedition had a memorable aftermath. In January 1976 the Royal Society staged a formal symposium to commemorate the first expedition to the Great Barrier Reef in the early 1930s, and despite my relative youth – for that institution in those times – I was invited to London to participate. The symposium was staged in the Great Hall, with paintings of British scientists of old leering down and accompanied by much pomp and ceremony. Incredibly, all but one of the original expedition members were still alive and able to meet us. The symposium was chaired by Sir Maurice Yonge (1899–1986), the expedition's famous leader and one of the first scientists to undertake rigorous experiments on reef organisms in the field.[11] Two days were spent presenting and discussing the results of our work, Stoddart himself being the man of the moment. The day before the symposium began we embarked on what could only be described as a pub crawl around London, with him spouting profanities all the way. I was again fascinated by his sparkling eloquence; he had one language when with the likes of me and quite another when VIPs showed up, at which point his choice of words was impeccable.

With that encouragement, when it came time for me to speak I decided to spice the tone up a bit by including my slide of David in his birthday suit on his last day of the expedition. This was met with stony silence. Not a single chuckle. I took a deep breath and went on. When I finished my talk I walked off the stage amid a decidedly muted ovation. Then, just audibly, I heard a deep voice in best Cambridge accent coming from the front row: 'Veron, you fucking little cunt, I'll get you for that.'

Things didn't improve. That evening there was a formal dinner, a *very* formal dinner, with about a hundred guests all dressed to the nines. I, resplendent in hired bow tie and dinner jacket, was seated in the furthest corner of the dining room, as befitted the lowliest of the company. Sir Maurice and David, along with the most important members of the original expedition, were seated at the high table on a dais. The wine waiter, a stooping old man in tails who looked as if he'd just emerged from a Charles Dickens novel, purposely poured just two fingers of wine into my glass.

'Aren't you a snotty old fart,' said I, upholding the best of Australian tradition.

Next round I received one finger. That meant war. At the end of the meal the old man handed me a decanter of port, as ancient protocol decreed, and glaring at me he told me to fill my glass and be so good as to pass the decanter to my *left*. He pointed left to aid my comprehension. Apparently passing it to the right, or anywhere else, was secret code for 'there's a traitor in the house'.

I looked him in the eye. 'That's not 'ow we do it down under, mate,' I said, and got up and walked around the table, in the wrong direction, filling everybody's glass.

That got silent stares from the high table. I thought Sir Maurice would have a stroke, but apparently he wasn't as bothered as I imagined. When he was invited to visit AIMS some years later I asked him and his wife Phyllis to dinner. I expected that to be a stiff occasion but it turned out to be quite the opposite. Sir Maurice regaled us with stories of his life, including, with a big grin at me, mention of a time when the Royal Society had a strange custom about passing the port. That night he drank so much wine I ended up carrying him to bed.

When Sir Maurice was a very old man, Kirsty and I visited him and Phyl, as she insisted we call her, at their lovely old stone house in Edinburgh. There was an ulterior motive for my visit, at least as far as AIMS was concerned. Among the many things my fledgling institution craved was old-fashioned respectability. Wouldn't it be special if AIMS acquired Sir Maurice's library, reputed to contain many old and rare volumes, and made a display of them at the entrance to its own library?

Just about the last topic I wanted to broach with Sir Maurice was what he intended to do with his books after he died, but he made the task simple by bringing up the subject of his pension. Not being money-minded – something I could relate to – he had retired on a fixed income, and of course that had gradually eroded. Now he had good reason to worry about Phyl's security, she being much younger than he. So he jumped at the idea of AIMS buying the whole collection, despite my protestations that it would be like selling part of himself.

And so the library was catalogued, Blackwell Publishing were assigned the task of valuing it, and AIMS promptly paid up, the amount being somewhat irrelevant to them.

'Disgraceful!' said David Stoddart, a dedicated bibliophile himself. 'Sending such a collection to that uncouth rabble of yours! Who's going to read any of it, eh?'

Well, I did, at least some of the old monographs, mostly because they were as much works of art as the science of the time, and I made good use of the volumes about corals. However, nobody else ever turned a single page of the most precious volumes. I know that because I kept the key to the glass cabinet they were housed in and nobody asked for it. Thus it was me who noticed one hot humid day that mould had invaded the cabinet. I made such a fuss about this that the whole collection was transferred, cabinet and all, to a museum in town where the ancient volumes would be properly cared for.

Stoddart, as only he could, turned the whole matter into his singular style of amusement. For the next couple of decades, I received the occasional letter from him on the same subject. 'My dear fellow,' they would always start, 'did you know that an original Grosse 1860 (a beautiful, long-forgotten volume) is up for sale?'[12] It has Sir Maurice's name in it; somehow it seems to have missed the AIMS sale. Such a shame.'

When David died in December 2014 I wrote a brief account of our friendship to be read at his funeral, one that I think would have made the lovable old rogue laugh again.

A Wayward Career

The first AIMS scientist

One evening the year after the Stoddart expedition, I had a call from Ian Croll, owner of a public aquarium on Magnetic Island, just opposite Townsville. Would I like to come out on his speedboat and try to find the *Yongala*? The SS *Yongala* was a 3700-ton passenger freighter that sank with all hands in a gale near Townsville in 1911. She is now a popular tourist dive spot, but in the '70s was almost unknown because she was hard to locate. Ian and I found her with the help of a spotter plane, and thus were among the first scuba divers to see her. She'd been plundered long ago, presumably by scavengers using pearl diver's equipment, but as compensation there were a couple of big groupers stationed at her bow, facing the current. They were so unconcerned about our presence that I put my arms around one for a photo.

On our return journey that day the weather turned bad, so it was late on a very dark night when we finally reached the beach opposite Ian's house on Magnetic Island. I jumped off the bow to fend the boat off the sand but found myself in a couple of metres of water. When I surfaced, spluttering, someone shone a torch in my face.

'Charlie Veron,' a woman said, 'what are you doing here?'

It was Isobel Bennett again. How did she always know where to find me? We agreed to meet the next day for lunch at Ian's place. When we did, Issie insisted I apply for a job at the Australian Institute of Marine Science, soon to start up in Townsville. I'd already thought about that of course, but imagined the place would not be at all to my liking as I was a self-motivated field worker, not the right person for the high-powered laboratories AIMS would probably have. Issie bullied me unmercifully and so I promised to write the letter. But I never did, because a couple of days later, Ken Back and his wife Pat invited Kirsty and me to a buffet dinner at their house to welcome Malvern ('Red') Gilmartin, the newly appointed director of AIMS, to Australia. Red, a red-headed stocky ball of American energy, had come to the Antipodes because he thought he would like the rough-and-tumble of Australians. Ken and Pat must have had fifty guests at their party, so I saw little or nothing of Red that night, and as I'm not at ease with such high-flying company I fortified myself with some of Ken's nice wine. In fact I became so fortified that according to one account I ended up dancing on his dining table. All I remembered was being the last person to leave and taken to task by Kirsty. She said my behaviour had been disgraceful. Red apparently had another view, which was that he'd just met the sort of Aussie he'd been told about and one who worked on corals to boot.

The phone rang very early the following morning. 'Are you Veron?'

'I suppose so, not sure.' My head ached horribly.

'What would you say if I was to offer you a job?'

'Dunno. What job? Who are you?'

'Gilmartin. To catalogue the corals of the Great Barrier Reef. Starting tomorrow. Your lab will be the large steel shed in the old quarantine station at Pallarenda. See you there.' He hung up.

Tomorrow was a Sunday. Nevertheless I followed Red's directions and there was indeed a shed in the old quarantine station, which I found after a little reconnoitring around Pallarenda, a satellite suburb on the north side of Townsville. With Muggin our spaniel in tow, I parked on the lawn next to the shed, which was locked. Then a man in a white lab coat came running up, screaming about trespassing on a quarantine station and telling me he was calling the Commonwealth Police, adding that I would get thirteen years for this. Poor guy, nobody had told him that his station was now the Australian Institute of Marine Science. I wasn't too sure about that myself, so I told him what he could do with his quarantine station and made a hasty retreat.

A few days later I received an official letter of appointment from Red, saying that I would be employed on a three-year contract to compile a taxonomic monograph on the corals of the Great Barrier Reef. The obviously multi-authored job description was divorced from reality, especially the time the work would take, but no matter. A week after that I received another letter, from George Melville, the general manager of AIMS, informing me that the director could not make appointments but that he could. The job description was exactly the same. A week after that yet another letter arrived, this one from the chairman of the AIMS council informing me that the general manager could not make appointments but that he could. The job description was again the same. That was my introduction to bureaucratic thinking, but I couldn't have cared less. I autographed a piece of paper and so became the first scientist to be employed by AIMS.

That meant resigning from James Cook University. When it came time for me to leave, Professor Burdon-Jones, who hadn't spoken to me since the incident on the *Marco Polo*, was a changed person.

'John, you are a credit to the university,' he said – he was just about the only person alive who still called me John. 'Your appointment has been one of the most satisfactory I have ever made.'

Really? That was an about-face if ever there was one.

'It's going to be difficult to find a replacement for you, especially at such short notice,' he went on.

'I know just the person,' I said. 'Terry Done. He's just finishing his PhD and was one of our mob at the Solitary Islands. Nobody better.'

'Well, John, I certainly respect your advice,' said the professor.

In later years Burdon-Jones referred to me in terms of ever-increasing praise and finally, after his wife and companion since childhood died, we talked a lot about his life. Despite our age difference and history, we became good friends. Strange how these things can turn out.

The origin of AIMS is unrecorded history, but I well remember many discussions about it. In 1968 Burdon-Jones had proposed to the Australian Academy of Science that an institute be established at James Cook University to provide the science needed for The Reef's future conservation and management. The academy apparently thought this a good idea, but found his plans too modest. By that time Australia had become a world leader in radio astronomy; what could Australia do next? With its long coastline extending from the tropics to Antarctic waters, the Pacific on one side and the Indian Ocean on the other, Australia was, they decided, uniquely placed for marine science.

The *Australian Institute of Marine Science Act* was passed in 1970, and after a prolonged delay an interim council (board of directors) was set up, chaired by Max Day. That was an extraordinary coincidence for me because Max had been head of CSIRO's Division of Entomology in Canberra when I gave my prize-winning paper there. Not only that, his personal research field at that time was colour change in insects. He knew my work well.

When I arrived at the quarantine station that no longer was, Red Gilmartin had brought an assistant over from America, and soon AIMS acquired a personnel officer, a purchasing officer and an accountant. It didn't take long for me to realise that battlelines were being drawn between Red and George. Mostly Red's fault, according to Max, who can't have been impressed when Red didn't show up at some council meetings and walked out on others.

The first thing I did was secure Len Zell, one of the founding members of our old scuba club, as an assistant. And so, with Terry Done lined up to take over my post-doc at the university and Len now assisting me, the coral part of our group at the Solitary Islands was getting back into action – on the Great Barrier Reef.

Nevertheless, I can't say I was overly thrilled with my new job. Our steel shed wasn't exactly welcoming and the whole place was rather empty, a far cry from my cosy spot at James Cook, where Kirsty and I had made close friends.

It was Burdon-Jones who suggested I stay on at the university until Terry arrived, and as that was going to be the best part of a year away I gratefully accepted. By the time I finally cleared my desk and moved to AIMS, quite a few staff had moved in, mostly imports from other countries, a colourful outgoing bunch.

A little before he was due to arrive, Terry asked Kirsty and me to reconnoitre the country around Townsville for a rural block of land where he and his wife might build a house. They changed their minds about this, but we went out reconnoitring anyway, and that's how we came across land for sale about ten minutes' drive south of the town, in a rural subdivision called Oak Valley. There were no oaks there, just scrub – mostly chinee apple, lantana and eucalypts. We found a superb 2-hectare block that commanded an excellent view of a tree-laden river and was completely private. Kirsty and I just looked at each other and thought the same thing: why live in town when we could be out here?

Within hours we had decided to build a house on the block and move out of our fibro box as soon as possible. We weren't going to be put off by trivial details either: the only access road was via a rocky ford across a creek that flooded every wet season, and there was no water, no electricity and no phone. Noni was thrilled at the prospect of living there, and after looking around for a while hammered a stake into the ground some 10 metres from the river and announced that this was where our front door should be. And that is where it is.

Russians for starters

Having accepted my job offer – all three of them – my first day on AIMS's payroll wasn't spent learning the ropes, but with Len on board RV *Kallisto*, a Russian research vessel out of Vladivostok. This came about because the Russians had applied for permission to undertake marine research in Papua New Guinea, at that time a protectorate of Australia. Red told me that the Australian government insisted on an Australian overseeing the ship whenever it was in PNG waters. The *Kallisto*, according to Red, could be a spy ship. Nobody else contacted me about the matter, which seemed a little weird, especially when Len and I received bright red diplomatic passports in the mail.

The ship was a rundown 4000-ton ex-German freighter, part of World War II reparations to Russia. Me in charge of this? I hadn't a clue what a ship this size did and I'd never even been to Papua New Guinea. If it *was* a spy ship, who or what was it spying on, and what was I supposed to do about it? What was I going to say to the Russian who actually was in charge? 'Sorry mate, I've got your job now'? These were the thoughts going around my head when I was taken to the cabin of the real scientist-in-charge.

Contrary to all expectations, Boris Preobrazhensky was delighted to meet me. He was a few years older than me, about my size, with a lively smiling face and an aura of warmth and goodwill. I found his broken English difficult to follow, but there was an interpreter aboard and with her help I discovered that Boris didn't expect me to be in charge of anything, that he wasn't in charge of anything either, in fact nobody was in charge of anything – which was how Russians normally did things.

Interpreter in tow, Boris took me around the ship introducing me to one person after another, mostly men, all of whom gave me a warm Russian-style welcome – a big hug and kisses on both cheeks. Being hugged and kissed by men was something new to me, but I soon got over it.

I had expected the Russians to be severe, humourless people, like they were on television. How wrong could I be. Almost everything that happened was turned into a joke, usually self-deprecating or taking the piss out of someone. This was apparently their way, at least outside Russia. It was very like the humour of my student days.

When we'd finished our tour, Boris turned to me with a faux-serious frown on his face. 'Only one rule important,' he said emphatically. 'You must repetition after my every English mistake. Russians very silly – make many mistake!'

'Russians are very silly – they make many mistakes,' I echoed. 'Silly buggers,' I added.

'Wonderful!' Boris laughed. 'Boris silly bugger!' From that moment on my name for him was Boris Silly Bugger, or just Boris SB. Thus it remained until the end of his days.

Boris, like most Russians who learn English at school, had a good vocabulary but he muddled the small words, especially articles, which don't exist in Russian. There was no way I could explain English grammar to anybody because I hardly knew any myself, another legacy of my non-existent school education. No matter, Boris and I talked a great deal and I corrected him automatically, even during deeply personal conversations. In no time his English was excellent.

Another interesting character on board was academician Yuri Sorokin, a well-known coral physiologist about twice my age. We found Yuri on the deck talking to himself. He grinned at me, showing a row of shiny metal teeth, and told me in easy English that he didn't speak English. He seemed quite mad, and indeed turned out to be appealingly eccentric and a delightful person.

That evening at dinner I was introduced to an attractive woman by the name of Galena who was studying crabs. She looked me up and down very slowly with a serious, intense look on her face and then murmured something in Russian that made everybody shriek with laughter. Boris later told me that she'd said, 'Isn't he gorgeous,' which immediately resulted in a new nickname for me: Gorgeous.

Galena's English was good enough for me to enjoy this sort of banter with her, although one day she looked completely blank when I asked her, just to amuse Len, if she often caught crabs. The ever-observant Boris noticed that something was going on and asked me what it was.

'Crabs, Boris – crutch crickets?'

He didn't follow. I tried again.

'Crabs – pubic lice?' He still didn't follow. 'As picked up in brothels!'

When the coin dropped, Boris burst out laughing and of course translated my joke for all to enjoy. Galena pretended to be shocked, but from then on, when the plan of the day was announced through the ship's public address system – as it was at seven sharp every morning – it always ended with, in English, 'And Gorgeous Charlie will catch crabs with Galena.'

One day, this actually happened. We had reached Conflict Atoll, at the far eastern limit of Papua New Guinean waters, and I was walking along the beach after a day's diving when I came across the biggest crab I'd ever laid eyes on. Unbeknownst to me it was the now rare coconut crab, the biggest of all land arthropods. How could I catch it for Galena? It looked capable of chomping my arm off. I was dressed in brief swimming togs and nothing else. I tried to get it to latch onto a piece of coral rubble. No luck. Then *Kallisto*'s hurry-up siren shrieked. As a last resort I took my swimmers off and caught a claw of the crab in a loop, then hurried back to the ship's company carrying the struggling crab in one hand and holding a piece of dead palm frond where my swimmers used to be in the other.

I never did live that down, especially once Galena embellished the story with her unflattering observations about my private anatomy.

The ship's chief diver was a giant who'd welcomed me aboard with a bone-cracking bear hug then announced that I was 'his little Australian'. He dragged me off to get fitted out with scuba gear, which turned out to be World War II junk, probably more war reparations: rusting cylinders, twin hose regulators, and what looked like buoyancy vests but which didn't provide any buoyancy. I spent most of my first dive getting used to the regulator, or rather trying to figure out how it worked.

Everything went well after that until one day, at a depth of about 40 metres somewhere in the midst of the Louisiade Archipelago, my regulator stopped working. I spent precious seconds trying to get it going by sucking and blowing. Making a free ascent from that depth was out of the question, so I took my tank off and pulled the regulator apart with the aid of my knife. After what seemed an age but was probably only half a minute, I was able to make air discharge through the demand valve by pressing a little lever with my finger. I couldn't use the mouthpiece because

I'd wrecked it, so I made a very uncomfortable ascent, breathing in mouthfuls of bubbles as best I could.

This created much consternation back on the *Kallisto*, and when the chief diver was summoned he made a brief inspection of the remains of the regulator and then angrily smashed what was left of it with his large fist. From then on I was to use his personal gear.

That evening the chief diver decided to make it up to me by insisting Len and I come to the ship's bar to drink vodka with him. The vodka, judging from its smell, was straight laboratory ethanol. The idea was that one shouted *Bud' zdorov* ('Be healthy', of all things) before tossing a beaker-full down the hatch in one gulp. This had to be followed by another gulp of 'cognac chaser', which seemed to be more laboratory ethanol mixed with brown sugar. The chief diver, with whatever company he managed to hijack, could keep this routine up for hours. I could hardly walk after just one round.

Boris had told me there were several KGB agents aboard, and warned me to be careful about what I said when they were near. I scoffed at that, telling him I came from a free country where people said whatever they bloody well liked.

'Don't be silly,' said Boris sternly. 'If you say something political they will write it down. They will make a long report about you. When you ask to come to Russia it will be *nyet*. That is how we live. All of us have a wife or children who are watched and would be arrested if we tried to escape. Make all the jokes you want, but keep your mouth shut about politics. You want to see my country, don't you?'

Point made.

One of the rather obvious KGB agents aboard was the ship's captain, a diminutive, neatly dressed man who clearly knew nothing about ships. He never allowed the *Kallisto* to come within a mile of a reef and never let her near water shallow enough to anchor. He seldom came down from the top deck, and as only the crew were allowed up, we seldom saw anything of him. One day that changed – someone told him that I'd played a game of chess with one of the crew. That sent him scurrying down to the back deck where we were preparing for a dive.

He knew only a few words of English. 'Chess?' he said. 'Now?'

'No,' I said, pointing at a pile of scuba gear. 'Diving.'

This routine – 'Chess? Now?' – was repeated later that day, then every other day, at any time. Apparently playing chess was the only thing the captain did. I gave in after a week of pestering and agreed to come to his cabin after dinner. The captain was waiting behind a beautiful chess set made of polished marble laid out on a low table. He offered me a glass of vodka. I declined, motioning him to have it. He declined. I had first move.

'Uh,' he said, grinning at me, 'you lose.'

He repeated 'you lose' with a grin after almost every move I made. Then he started to tap his fountain pen on the table every time it was my move. I pointed to his hand and motioned him to stop. He did, for a move or two, then started up again, *tap, tap, tap*, 'You lose.' Again I asked him to stop. He had taken control of the game and was clearly winning. Every move was broadcast over the ship's intercom, which I also found a tad unnerving. *Tap, tap, tap*. I reached over, took the pen, broke it in half and handed it back to him with a big smile.

He was furious. He glared at me, clearly trying to decide what to do about it. He made an angry move, slamming a piece down on the board, then another. Both mistakes; now *I* had control of the game. The captain recovered, started smiling again, and also started winning again. Luckily for me he was too late; I forced a draw.

Although it was well after midnight, several officers burst into the cabin as soon as the game was over, and there was much chatter. The captain turned the board around and promptly started a new game. I should have left when I was half ahead but instead I threw in the towel after only a few moves. The captain was ecstatic; apparently Western decadents couldn't play chess. Every day, for the rest of the trip, he would venture down to the lower decks to grin at me and say, 'Chess? Now?'

One morning the intercom announced that we were abeam of the mouth of the Sepik River and a shore party was being organised. Len and I hurried to the longboat; there were about six of us. As we neared the shore I started to have misgivings; wasn't the Sepik home to cannibals famous for head-shrinking? A memory of Melbourne Ward's shrunken head troubled me.

The beach was deserted but behind it there were several thatched huts on stilts. This was enough for Yuri Sorokin; as soon as we landed he scurried up the beach and disappeared into the dense jungle. A few women appeared, then some children, still getting their clothes in order. I relaxed. An hour or so later drums started up, and then a band of men arrived, dancing in file, all dressed in full ceremonial regalia – painted, covered with bird plumes, with big scary masks and long penis gourds. One of the men, dressed just like the others, was Yuri Sorokin. After that we all joined in.

The Russians were delighted that the Sepik natives had Russian words in their vocabulary. These had come from the Russian Lutheran missionary Nikolai Maclay, who lived there in the late nineteenth century. Papuans are renowned for their language skills; I could see why.

I made unforgettable friends on the *Kallisto*. Boris especially, but also Yuri, Galena, some of the other biologists, and several of the ship's crew. The day we parted company, Boris, by then speaking almost fluent English, was unusually serious. He said, 'Charlie, there are many differences between our countries and it could get worse. The greater the differences get, the stronger our friendship must be.'

Wise words: our friendship outlasted the Cold War and throughout all that time we kept in contact. Of course we promised each other reunions, in Australia and in Russia, both of which happened.

Shortly after returning to my lab at James Cook, a man dressed in a suit and tie and carrying a briefcase knocked at my door and asked if I could help with his inquiries. He said he was from the Australian Security and Intelligence Organisation (ASIO) and had just interviewed my father and both his neighbours; furthermore he knew that I'd been involved in anti-Vietnam War demonstrations as a student. I'd heard nothing from ASIO before my trip, but now this man told me that my association with Russians was 'of concern to Australia'.

I told him a little about the *Kallisto* and said I'd seen no evidence of it being a spy ship.

'Of course we expect you to say that,' the man said.

What?

'We believe that you may have been recruited,' he went on in the most offensive way, and said it was his job to find the likes of me, who always ended up spending a long time in prison.

Recruited to spy on my own wonderful and utterly loved country? I exploded. I grabbed him and threw him out of my lab and down the stairwell, just across the corridor, then hurled his briefcase at him. It missed – I was always a rotten shot – and broke open against the stairwell wall, pens and paper flying everywhere. He shouted threats at me, and I back at him. It was a hell of a row. People came running from all directions and some started shouting at him also.

I heard no more about this incident in Australia, but as I was soon to discover, America took a different view.

Boris's trip to Australia the following year, 1976, wasn't difficult to organise. AIMS had money for visitors, and being a government institution could cut through the red tape, which was thick at the time for Russians. He arrived from Moscow and I met him in Sydney. We travelled around a little, then went on to Canberra, and finally up north to stay with my family.

Noni was very taken with Boris, which was unusual for her. Most adults who didn't know her treated her as a child, which of course she was, but that usually meant talking down to her, a great mistake. Not so Boris, who was soon in the river next to our house, covered with water lilies – a river monster out to get her. I have vivid memories of him reading Russian bedtime stories to her from a beautiful book he'd brought from his home.

Just before his return to Russia, I asked Boris what he thought the biggest differences were between his country and mine. 'The expression on the faces of women,' he replied without a moment's hesitation. Nothing about houses or cars, or the news media attacking politicians. I thought his reply very strange.

It wasn't until 1979 that I found out why. I was invited by the Russian Academy of Sciences to attend a symposium in Khabarovsk, in far eastern Russia, as a VIP. Me a VIP? That was certainly Boris's doing. However, it meant that I wouldn't have to travel under the auspices of Intourist, as all other visitors to Russia were obliged to do, and there was a special flight from Tokyo for foreign delegates to the symposium as there were no commercial flights to eastern Russia. The airport guards I encountered were horrible people who shouted orders at us; then I heard Boris shouting orders at them from somewhere in the airport building.

True to my nature, I did some very silly things on that trip, smuggling in a whole crate of mostly forbidden books (the crate being labelled 'scientific specimens'), smuggling out a forbidden manuscript, and travelling up the Chinese border to the city of Amur and then on to more places where all foreigners were banned. As long as I kept my mouth shut and left all the talking to Boris, I thought I looked exactly like a Russian, but this wasn't so. Twice I was recognised by villagers as a foreigner and had to abandon my disguise. How did they know? By the way I walked, they both said; apparently people who have spent their lives in fear recognise such things.

I've always loved journeys like those I had with the Russians, getting to know local people in ways most foreigners never could. It's a great privilege. Russia was a harsh country, especially for women: I don't think I saw one smile the whole time I was there, just as Boris had said. Yet the real people of Russia were nothing like those I'd seen on television in Australia. Were it not for politics, Australia and Russia could have been sister countries.

Boris Preobrazhensky and me on the RV *Kallisto*, 1975.

Big dusty museums

My work, to 'monograph the corals of the Great Barrier Reef', could no longer be described as population ecology; it was taxonomy, and that meant mastering the ancient monographs and type specimens – or holotypes, the specimens on which names are based – in museums. The British Museum (Natural History) in London, the National Museum of Natural History in Paris, the Museum für Naturkunde in Berlin, and the Smithsonian Institution in Washington DC were the world's main repositories of these specimens, including most of those from the Great Barrier Reef. These were places I had to not only visit, but spend much time in, longer than my sojourn on the *Kallisto*.

Kirsty and I discussed this at length; should she come with me? What about Noni? She was too young to get much out of an overseas trip. It was a horrible thought, but clearly the best thing would be for her to stay in Australia, with Kirsty's mother or mine. She would be safe and happy with either. At that point Kirsty came up with another idea: why not send Noni to her mother's and have my mother come with us? She had never left Australia and would be thrilled to go travelling around Scotland and France with Kirsty.

Mothers-in-law are supposed to be perennial problems for their daughters-in-law, the family member who must be tolerated. I joked about this, because Kirsty and my mother were the closest of friends; they needed no excuse to seek out each other's company and were always chatting about whatever it was we were planning. Not surprisingly, my mother jumped at the idea of seeing the world for the first time with Kirsty; it would be her trip of a lifetime. Kirsty's mother was more than happy to have Noni, and Noni thought the idea of a holiday with her Mackenzie gran, at Stonehenge of all places, was as good as it got. And so, after just a few phone calls, all was arranged. Noni would go to Kirsty's mother; Kirsty, my mother and I would go to London; I would work at the Natural History Museum, and Kirsty and my mother would go on tour.

On 22 June 1975, only a couple of weeks after these plans had been made, I came home from a short diving trip to find Len there. On seeing me he hurried past with a pained look on his face. I found Kirsty crying her heart out. She hugged me close and told me my mother was dying.

My father had been driving down the Pacific Highway in Sydney with my mother next to him when a drunk youth, racing in the opposite direction, lost control of his car and crashed head-on into Dad's. Although both cars were wrecks, Dad wasn't injured, and at first he thought Mum was also okay. But her neck had been broken and in hospital she was pronounced dead within a couple of hours. The driver of the other car had been knocked unconscious, still with a bottle of whisky in his hand, but was uninjured.

I was confused. I understood what had happened, but I couldn't *feel* it had happened. My mother had always been there for me, every day of my childhood, then every day of my school life. She was my anchor, my guide, my fountain of wisdom. It was beyond me to realise that all of this, my lifelong blanket of unquestioned love and security, had suddenly ceased to exist.

My thoughts for my father, sister and grandmother came to my rescue; I plunged into worrying about them rather than myself. I believe to this day that my mother and father had a perfect marriage, an absolute love affair going back to the time when my mother had only just left school. If they ever disagreed about anything, and surely that must have sometimes happened, I never noticed. If my mother wanted something done, Dad did it. I knew her death would be a catastrophe for him. I just wanted to be with him, and with Jan, for whom it must have been agonisingly real as she was still living in the family home. I tried to imagine how it would be for my grandmother, who had lost her only, much loved daughter.

Mum's funeral was at St Alban's, the church where I'd spent so much time as a child. I will never forget that day. Not only was the church packed, standing room only, but the surrounding grounds were also. There were hundreds of people I barely recognised;

I'd had no idea how big my mother's world was. It seemed that half the shop owners of Lindfield were there. Through the mist of my grief I saw our Italian greengrocer with tears pouring down his face, and our milkman, and postman, and so many others I didn't know. That day when I said farewell to Mary Veron was the day I discovered who she really was.

At that time I also discovered something unexpected about Noni. I would have thought that a five-year-old wouldn't understand death, especially one who'd never come across death before, as far as she could remember. Yet Noni seemed to understand completely. She knew she would never see her gran again and she would not be consoled. Gran was the most wonderful person in her world and her closest friend. For years she clung to the last present my mother had given her – a plastic, silver-painted clockwork bell that played 'Jingle Bells'.

The loss of her gran had a permanent effect on Noni, giving her a head far older than her age and leading her to think deeply about the big things in life – good and evil, and who she herself was, or wanted to be.

My mother had died only six weeks before the three of us were due to depart for England. I felt at first that I couldn't leave my father, but as time wore on it became clear that there was nothing I could do for him. He had lived with death throughout the war years and perhaps this had toughened him, I don't know. My sister was with him, now his faithful housekeeper. My grandmother suffered alone. True to her stoic, Victorian, impenetrably deep nature, she kept her tears to herself, but I knew that facade well, we were very close.

Back home, still in a haze, I realised that if I didn't work on those museum specimens now, I would have to do it some other time, and soon. Kirsty knew this also, so we decided to go ahead with our plans, except for the touring, which Kirsty said she couldn't face alone.

I had never been to a foreign country at that point, apart from Papua New Guinea, and Kirsty had only been to New Zealand on holiday as a child. In fact, I had hardly seen a building much older than my grandmother, and perhaps for this reason I'd never taken any interest in history, which I loathed at school along with almost everything else. And so I had a strange feeling as our plane circled down over London and I saw the houses of parliament, the Tower and Big Ben. Of course I knew they existed, but the old saying 'seeing is believing' became all too true for me. That feeling of profound ignorance of something important took a leap forward when we did a bus tour around London and ended up at Windsor Castle. Who built this place? Why? Who paid for it? How did the plumbing work? I had questions without end. Kirsty could answer many of them as she'd majored in history at university, but more and more I felt I had to learn about history myself. And so I started a personal journey of discovery of the history of England, vaguely like my discovery of classical music. I developed a fascination with the past – piecing it together – which has never left me.

At first I was taken aback by the corals in the British Museum because there were so many I'd never seen, but those from the Great Barrier Reef, of which there were thousands, all had a feeling of familiarity. I started to recognise the insights and mistakes of taxonomists of the past, especially when they'd given a new name to what was only a variation in growth form due to the environment in which the coral had grown. I made seemingly endless notes. It was tedious work, as there was sparse information with most specimens, usually only a label saying who'd collected it and roughly where it was from. For many there wasn't even that. Not surprisingly I became interested in the personal side of the collectors, in fact I had to, to make sense of what they wrote.

Gradually, too, I learned to interpret old manuscripts in ways that approached reality as their authors saw it. It was Henry Bernard who interested me most, because he described so many species and had much to say about most of them. Like several other taxonomists of his time, he never saw a living coral and, taking tunnel vision to an extreme, thought there was no reason to do so. And so he had little understanding of the corals he devoted most of his career to writing about. (He also went off on odd tangents, like criticising Darwin's theory of atoll formation.)

A type specimen displayed in the British Museum, borrowed from Paris's National Museum of Natural History.

What concerned me most was an almost complete lack of study by Bernard and his contemporaries of anything but single specimens, or at most small groups of specimens. Why describe a coral as a new species without knowing, or even wanting to know, how it differed from other species? Bernard's work was mostly highly respected by his peers at the time, but they might have been forgiven for considering it pointless, as some ultimately did.

For two long months I battled corals in that dusty museum while Kirsty occupied her time visiting places of interest and revelling in music concerts and the theatre. When I'd had all I could take we hired a car and made the tour of Scotland that Kirsty had planned with Mum, staying in B&Bs or with the families of our Townsville friends. But we both longed to be back with Noni. I was very envious when, after a trip to the Netherlands, the time came for Kirsty to head home. There were more museums on my itinerary, in Europe and America.

At the museum in Paris I was sidelined somewhat by the French for not speaking their language, but the museum's coral palaeontologist, Jean-Pierre Chevalier, came to my rescue and we spent a good deal of fruitful time finding type specimens that had been presumed lost or were mislabelled. I wrote notes to go with these, tying them to the specimens, but saw during a later trip that they'd all been removed. No surprises there: how dare a novice, and an uncouth Australian at that, attach notes to such historic specimens.

From Paris I went to several other European museums, the most interesting being Berlin's, on the Russian side of the Berlin Wall. I was refused permission to stay overnight in the east, so every morning I had to go through Checkpoint Charlie, which turned out to be well named for I was always searched and questioned by Russian-speaking lowlife called border guards, just like those at Khabarovsk. However, in the museum it was a different matter: the Germans were delightful people and I made a friend of Dieter Kühlmann, one of the museum's curators, who years later was to join me in Japan. I found many ancient type specimens from the Red Sea, some still in packing cases in the basement where they had been put for safekeeping during World War II. Many of these were of great historic value, especially the type specimens described by the German coral taxonomist Christian Ehrenberg in 1834, which I soon realised could be distinguished from the others in his collections by a black border he'd drawn around their labels.

After Germany it was on to America, where I was warmly welcomed by one museum after another. I spent most of my time at the Smithsonian, where most of the types were kept, but not surprisingly I had the same problems with the corals as I'd had in Europe. I found many were as pointless as those in the British and Paris museums because new species were described from single specimens and had little associated information. But others were not: I was especially impressed by the work of James Dana (1813–1895), a geologist who took up corals as a sideline. He'd had insights into coral taxonomy that none of his contemporaries remotely matched, mostly because he clearly did his own field work.

At the other extreme was A.E. Verrill (1839–1927), who published an astounding amount of taxonomy, only a small part of which, fortunately, is about corals. I gained the impression that he was considered some sort of hero by Yale's Peabody Museum, but as far as I was concerned he seemed to have no idea about corals, apart from what other people had to say about them and from rummaging through the specimens others had collected. His holotypes were mostly barely identifiable fragments. Worse, he had a habit of depositing one holotype in one museum (usually the Peabody) and another of the same name, but actually a different species, in another museum (usually the Smithsonian). His accompanying description sometimes appeared to be of a third species. He also got type localities wrong: *Montipora* from the Gulf of California? *Platygyra* from Hawaii? I don't think so. Such a mess. Perhaps he did better with other groups of animals, but I can't imagine that he ever saw a living coral.

A few years ago I emailed a curator at Peabody about one of these problems and was surprised that he remembered me. To my delight had a name for the mess: Verrilliana.

My non-working time in America had an amusing side. On arrival I made a beeline for the Smithsonian, where a kindly curator who studied sea fans installed me in a nearby hotel frequented by museum visitors. That night, feeling a little jetlagged, I went for a walk. I usually head for a park or botanical garden when stuck in a city, but as they seemed in short supply around my hotel, I followed my feet to what appeared to be Washington DC's red-light district. There I found streets that kept me entertained for miles. Chatting about my walk at morning tea at the Smithsonian the following day, I noticed that people around me had stopped talking and were listening closely.

'Do you have *any* idea how dangerous it is where you went?' someone blurted out. Most of those in the tearoom declared they wouldn't even drive down some of the streets I'd walked, even in broad daylight. None of this had remotely occurred to me: that was the joy of being an Australian who thought danger was all about snakes, sharks and spiders, not people.

That evening, back in my hotel room, I noticed a row of locks on the door, a chain and a spy glass. In the lobby a sleepy clerk told me she wouldn't set foot outside the front door after dark under any conditions. I felt like I was in a prison.

Then I remembered I had a means of escape. Not long before this trip a friend of Kirsty's family, an old judge, promised to give me a phone number to call if I wanted a place to stay in DC. The judge died soon after this, but his wife remembered his promise and looked the number up for me in his diary. I was reluctant to make that call, but stuck in my hotel room I felt I had to do something.

A woman called Margaret answered the phone, and I explained who I was and wondered if she could put me up for a couple of days until I found a hotel in a less dangerous area. She gave me her address and told me to call a cab.

Her house was a very grand mansion. I was shown to a guestroom the size of an apartment, complete with walk-in cupboards, a bath that looked more like a swimming pool, even a private sauna.

I stayed with Margaret for my entire time in DC. She was middle-aged, very bored, incredibly wealthy, and so were all her friends, to whom I was introduced as 'the house guest from Australia'.

For a while the house guest amused himself at a succession of dinner parties by telling ever more incredulous stories about killer sharks the size of submarines, and kangaroos that delivered mail in their pouches. In return I was propositioned by one bored female after another, reminding me of the film *The Graduate*. But I had no wish to visit luxury yachts in the Bahamas, or villas in the Appalachians, I just wanted to work on corals. And so I was chauffeured to work in a big black Cadillac every morning and collected in the Cadillac every evening, usually to be dragged off to yet another party and surrounded by yet more rich, bored, middle-aged women.

After three weeks my departure drew close. Over breakfast one morning I brought up the subject of the judge, who I had to thank for arranging my stay.

'What judge?' Margaret said, looking surprised. 'I don't know any judge.'

I had dialled the wrong number. An Australian accent worked wonders in America in those days.

My final stop was Cornell University, Ithaca, to see John Wells, a professor of geology. Considered by most to be the world's foremost coral taxonomist, John was an elderly yet sprightly character with a lively sense of humour. We hit it off immediately, especially as he was keen to hear my views on how corals varied in structure according to where they grew and the consequences of this for taxonomy. I chatted on about the possibility of creating a single global taxonomy of corals. Discouragingly, he thought this an impossible goal, however desirable. We also talked about coral distributions, he being the world's keeper of records.

John had two houses, one on Cornell's campus, his winter home, and the other on the edge of Lake Ontario. He called the latter house Lucky Stone, because of the many stones around it with holes in them. These, he explained, were fossil wormholes, although he didn't say why they were lucky. On his study wall at Lucky Stone, John had a large chart showing the distribution of coral genera. These were ranked in descending order, from most common to least, down the first column, and countries were ranked in descending order, from most diverse to least, across the top. The chart was covered with crosses in pencil that indicated a genus was present, and there were dozens of changes and question marks. Changes in the order of the rows or columns had resulted in multiple patches being stuck on with Sellotape, some of it ageing and peeling off.

I spent an hour or so adding new crosses here and there, and more question marks, all of which we discussed. John told me he'd redrawn the whole chart several times. Clearly a new edition was needed – this one was falling apart – but I was a little taken aback when, our discussions over, he ceremoniously took the chart off his wall, rolled it up and handed it to me.

'Here you go,' he said. 'Look after it, won't you.'

I felt humbled. I later helped myself to a lucky stone and hung it around my neck.

The following year, 1976, I was invited to a coral taxonomy workshop hosted by a marine station on Enewetak Atoll, in the Marshall Islands. It was an ideal opportunity, as those invited included many taxonomists I'd befriended during my time in museums. But my impulsive past came back to haunt me. The marine station was on a US military base, and the US apparently didn't approve of me throwing an ASIO agent down a stairwell. Permission to travel to the Marshall Islands was denied.

After a flurry of letters, I was eventually given a visa, valid from the exact day I arrived in Hawaii en route to Enewetak Atoll to the exact day I was to leave, and so it remained for all my trips to the US until the system became computerised. After that I was 'randomly selected' for searching on American flights – for twenty-eight flights in a row.

John Wells was the guest of honour at the Marshalls workshop because he had done much of his original work there. He'd been fiddling with photos I'd sent him of a highly variable species that he had named *Plesiastrea russelli*. I had changed both his species name and the genus to which he'd assigned it.

'Charlie, isn't this going a bit far?' he remonstrated. 'Some of these photos, surely, are of other species – several other species, I'd say.' The species in question was common around the islands.

'Okay John, let's take a look at some live specimens,' I said.

I fetched my scuba gear and collecting basket, dived down to about 50 metres, and gathered a series of specimens at regular intervals all the way up, finishing on a sunny, wave-hammered reef crest. I cleaned them and laid them out on a bench in sequence. John spent hours pottering up and down, looking at them with his hand lens, grouping them one way then another, as I looked on trying not to smile, for I had done something similar myself many times on the Great Barrier Reef.

'I surrender,' he said, straightening his back. 'Most palaeontologists would have called most of these specimens a different species, and they would probably have made several genera of them. But I agree, they are all one and the same species.'

By that time I was confident of my work, but the consequences of environmental variations in species for taxonomy was something still not accepted by most coral biologists, so it was good to have John's affirmation.

Rivendell

This very interesting trip was somewhat spoiled on my return home. There a letter from AIMS awaited, informing me that my contract to work on coral taxonomy had been changed from three years to two, and that I was to be reassigned to 'other work'. I was devastated, then outraged. Coral taxonomy had grown tremendously important to me, and so had Rivendell, my new country home. It would be impossible to keep Rivendell without a good job, and the only job I wanted was the one I had. My work was turning into something original and worthwhile by any standard. Nevertheless, I felt I had no choice but to act decisively: I resigned my job, cleared my desk and left.

It was a horrible time, especially as I knew I was putting my family's security at risk. Fortunately, the pain didn't last long: a few days later I received another letter, this time from the chairman of the council, referring to the institute's actions as a mistake and extending my contract to work on coral taxonomy to four years. I realised then that I had been caught up in one of the endless power struggles within AIMS. My semi-permanent war against bureaucratic nonsense had begun.

Something else had been waiting for me on my return: a package full of police documents from my father. He had found out that the teenager who was driving the car that killed my mother had only been charged with a misdemeanour. With a little digging, Dad had discovered that nobody in New South Wales had ever been charged for drunk driving leading to a fatality. Why not? Such accidents were happening every few weeks. With further digging, and help from some co-operative police, he'd found out that the going price for getting charges dropped was $18 000. This affected Dad badly. He was not so concerned that Mum's killer had got off – he was just a kid who'd been very silly – but he was furious at the thought that thousands of Australians of his generation had given their lives for their country in war only to have that country governed by corrupt officials. They, Dad said, were making a fortune out of road deaths.

In his strange way, Dad had been a crusader for justice as long as I could remember; now he was on a warpath. As the months rolled by, he repeatedly phoned me to talk about his latest accusations, many of which came from confidential files held at police headquarters in North Sydney. He'd been able to photocopy, even 'borrow', hundreds of files because straight cops would leave the key to the filing room on a counter and turn their backs when he arrived. He also received anonymous tipoffs. But even with these files, some of Dad's assertions seemed to me to be over the top. Was the then Premier of New South Wales, Robert Askin, really in on this? Was it really masterminded by the police commissioner himself? Dad would write what he called 'brochures', giving details of one case after another, and post them to everybody he could think of, from the Queen down. He was trying to get himself sued for libel – the only hope he had of getting his allegations into court. He was threatened repeatedly but never sued, evidence perhaps that he was on the right track but possibly that he was universally regarded as some sort of crank. One day he accused a high-ranking judge of being involved in millions of dollars' worth of police corruption. I found that unbelievable. I begged him to see a psychiatrist, and much to my surprise he did. The psychiatrist found nothing wrong with him. On he went; it was a complete obsession.

We moved into Rivendell in the winter of 1976, after the shell of our hideaway was finished. No electricity meant no running water, so we bathed in the river and cooked on a camp stove by lantern light. It was cold for the tropics, so we went to bed early to keep warm, listening to the wind in the trees, the croaking of frogs in the river, the guttural arguments of possums, and the mournful cry of stone curlews in the distance. Come dawn we would hear the splashing of cormorants as they fished in the river, the call of kookaburras and the chatter of parrots.

I found a secluded place down by the river – dark, enclosed by trees, peaceful. I would go there and sit on a tree root and not think about anything, at one with the river and with myself. A little later I'd emerge refreshed and renewed. It was a special place; it still is.

It was Kirsty who named our house Rivendell, from Tolkien's *Lord of the Rings*: 'A perfect house, whether you like food or sleep or story-telling or singing, or just sitting and thinking best, or a pleasant mixture of them all.' It was an idyllic life as far as we were concerned, although the list of jobs to be done was dreadfully long, especially the building of a driveway that would still be a driveway when the wet season arrived, and the clearing of chinee apple and lantana so that we could see the river from the house, not just hear it. Then there was a lot to do in the house itself, because we'd run out of money to pay our builder. Nevertheless, the first job I actually did was build Noni a treehouse. About 30 metres from our house there was a big eucalypt leaning out over a bend in the river, its branches affording a good view in both directions, and there was deep water beneath. I made a ladder, a platform and a little hut. When it was finished, Noni christened it Gum Leaf and insisted on sleeping in it alone the first night.

Being somewhat isolated at Rivendell, Noni often had to look to herself for company and entertainment. She became quite a daydreamer and frequently had her nose in a book, switched off from all else. Memories of my own childhood flooded back, especially when her interests and hobbies became entangled in her daydreams. Rivendell, books, and spending a lot of time in her own private world combined to turn her into another naturalist in the family. She was always lost in the delights of her own discoveries and this developed into a thirst for knowledge on all manner of subjects. Perhaps Kirsty and I influenced her a bit, but mostly her interests came from within, just as they had with me. We had

Long Reef rock platform at low tide. At the top is a golf course, then a little promontory (near the centre) that has excellent picnic spots. The big rock platform below the promontory is only exposed at low tide. My favourite collecting spot was at the very bottom.

John Rotar's 'big boat' at the Solitary Islands.

A reef flat exposed at low tide on the Great Barrier Reef. The corals seen here and in the photograph below are at environmental extremes and have little in common.

A garden of coral in a protected muddy embayment on the Great Barrier Reef.

A coral releasing egg and sperm bundles at night during a mass spawning event.

The undersurface of a large *Acropora* colony. The colony consists of hundreds of interlocked radiating ribs, each rib being built by about a hundred individual polyps.

Ribbon reefs. The outer face (right) plunges into the Queensland Trough. The reef front is wave-hammered for most of the year.

A heavily laden *James Kirby* heading north on its maiden voyage, the Stoddart Expedition of 1973. As of 2016 this amazing vessel was still in service.

The beacon on Raine Island built by convicts in 1844.

Len Zell

Turtles starting their journey up the beach at Raine Island.

Lord Howe Island lagoon, the southernmost coral reef in the world.

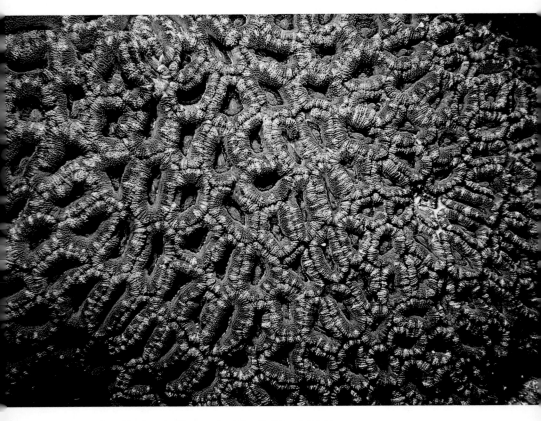

Acanthastrea lordhowensis at the Solitary Islands.

Clerke Reef, Rowley Shoals. The sandbar is to the left and lagoon to the right. The largest channel is great for a fun dive.

Sea gypsies' boat.

Spear fishing with wooden-framed goggles on Clerke Reef.

Coral community at the Houtman Abrolhos Islands of Western Australia. This mix of species is seldom seen in other coral reefs.

Acropora seriata at the Houtman Abrolhos Islands of Western Australia. The greenish clumps on the left and the plates beneath are different growth forms of the same species. This species is abundant at these islands but has not been recorded elsewhere in Australia.

Small plate corals battling with *Sargassum* on a Houtman Abrolhos reef slope.

Ishigaki Island, showing living corals in the foreground, the reef flat (yellowish), the beach (white), the coastal strip of agricultural plots, and the mountain ridge behind.

A coral knoll at the Ryukyu Islands, Japan. It is a formidable job identifying each and every coral.

Reef slope, Chumbe Island, Madagascar. This is a favourite photo of mine because it shows the many growth forms corals have.

The *Inga Viola* at top speed, Madagascar.

Coral community at Clipperton Atoll, built almost entirely of *Porites*.

Victoriano Álvarez's home inside the rock on Clipperton Atoll. The bright orange crabs that finally ate him fill crevices in the rock and are scattered over the sandy floor.

The first scleractinian corals (Middle Triassic). These have structures like corals of today, but did not all occur together at the same time and place.

Common growth forms of modern scleractinian corals.

Devonian reef, Canning Basin, Western Australia.

Late Triassic reefs of the Austrian Alps.

Goniopora pandoraensis with some branches bleached. This is the first photograph of bleaching on the Great Barrier Reef.

Azooxanthellate coral. The bright colour comes from the coral itself, as these corals have no zooxanthellae.

Bleached coral on the Great Barrier Reef.

A healthy upper reef slope.

A reef slope two years after a mass bleaching event. The corals still alive are being smothered with seaweed because of an ecological collapse.

A fanciful eighteenth-century painting of coral collecting.
Many coral names used today stem from these times.

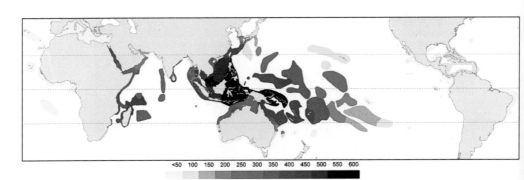

The global diversity of corals computed from Coral Geographic (coralsoftheworld.org).

long discussions about one subject after another – her subjects, not mine or her school's. We'd talk on until I found myself telling her things she hadn't asked about, and saw that she wasn't listening. How well I understood that.

Noni's love of animals followed her interest in Nature. Rivendell was the perfect place for a dog, so we bought a little puppy, a German shepherd-cross. He promptly moved into Noni's bedroom, and would have slept in her bed given half a chance. She also assembled an assortment of birds: geese, ducks by the dozen, and later a large turkey who could demolish an entire chilli bush in one go. She kept goldfish in her bedroom and frogs, tortoises, snakes and all sorts of caterpillars in Gum Leaf or around the house. One special pet was a children's python called Mundingburra, the Aboriginal name for her school. Seeing her with all these animals had me reliving my own childhood.

With our encouragement Noni began to tinker with the piano, and then she started lessons. This was of course very much Kirsty's domain, but I decided to motivate Noni with a race: which of us could learn the quicker? At first it was no contest; I picked up her pieces much faster than she, and so became a bit concerned that our little contest might discourage her. That worry didn't last long; I started to plateau out and she didn't. Within a year, awards started rolling in, then her teacher decided to enter her in the under-eight solo piano division of the 1977 Townsville Junior Eisteddfod. Her Beethoven sonatina was a nailbiting performance for Kirsty and me, but Noni was completely relaxed about it and she won. There was no stopping her after that. She needed neither encouragement nor reminders to practise, rather the opposite – we often had to nag her to stop playing and come to dinner while it was hot, or to get ready for bed.

There was always music in the house, either live from Noni or Kirsty or from my old stereo, still going. It was a good life.

Examining a gecko with Noni in Gum Leaf, her treehouse, 1976.

Monographs and bureaucracy

By this time I was installed in the steel shed at Pallarenda, where Len had been working for almost a year. He had erected rows of large particle-board shelves held up by concrete blocks, and a long bench perched atop of a string of 44-gallon drums. It didn't look too good but it was functional. We had thousands of coral specimens, all cleaned in tubs of chlorine bleach of the kind used in swimming pools, most of which had been collected during trips to the Palm Islands. We had established a makeshift camping area on Orpheus Island, at the northern end of the Palms, and later kept scuba tanks and a compressor in a shed there.

I started writing the first volume of the monograph series called *Scleractinia of Eastern Australia* in the steel shed. It had no insulation, let alone air conditioning, and was oppressively hot all summer. Len and I would be stripped to the waist, pouring with sweat, and I had to rest my writing hand on blotting paper lest sweat smudged my script. I eventually gave up and wrote the whole monograph series, and everything else I ever published, at Rivendell.

It was the AIMS council that decided my monographs would be a joint production between AIMS and James Cook University. I was never consulted about this, nor presumably was Michel Pichon, a Frenchman newly arrived from Madagascar, who was made a co-author. Michel contributed an account of *Psammocora*, an obscure genus of corals I knew little about. This helped fill the first volume of the series, which even so remained precariously slim. But I had to go ahead and produce it quickly as AIMS was anxious to have its first major publication on show.

The basic thrust of that first volume was my original notion that corals vary their growth form with environment, something that can only be studied by divers. The implication was that the taxonomy of museum workers of old was wrong, an idea that John Wells was starting to agree with. Not surprisingly, when this was published there were many objections. Some came from European palaeontologists funded by the petroleum industry, coral reefs being good indicators of underlying oil deposits. Pat Mather, still apparently outraged at my original appointment, wrote a furious letter to the Minister for Science saying that this was what happened when novices dabbled in taxonomy.

Don't trouble yourself too much, Pat, the coin will drop sooner or later.

Red Gilmartin dismissed all this with a wave of his hand, but called in a couple of extra reviewers 'just for appearances'. John Wells wrote that he found the volume 'refreshingly original'; David Stoddart said that it was 'returning coral taxonomy to the reality of the reef'. I don't know what Michel Pichon thought of these goings-on; I doubt that this was how things were done in France.

Maya Wijsman-Best, a Dutch coral taxonomist who had recently completed her PhD in New Caledonia, enthusiastically agreed to work with me on volume two, which would cover the next group of corals, her specialty. Maya proved to be a kindred spirit and accompanied me on several field trips. She also helped with the historical side of things, linking in the old monographs and formal taxonomic protocol. It was a good partnership and I was sorry when it ended. This volume, published in 1977, was a relatively unhurried affair and attracted good reviews, probably due to the range of corals it covered, which were less contentious because most didn't have elaborate growth forms.

This was also about the time that AIMS moved from Pallarenda to its new headquarters in remote bushland about forty minutes' drive south of Townsville. A lasting tribute to bureaucratic know-how, the choice of this site has to be one of the dumbest decisions in the history of Australian science, for it meant the entire staff had to spend over an hour driving to and from work every day, for no reason whatsoever. Furthermore, it ensured that AIMS was isolated from both the city and the university.

Many years later, when I was given the task of producing a history of the first twenty-five years of AIMS, I was determined to find out how this crazy decision had come about. It was all because three members of the council had hired a helicopter and thought the site, with its long beachfront, looked beautiful. And so it does, especially for those with helicopters. It's rather less attractive for those who must drive there and back every day.

My laboratory at AIMS: the shed at Cape Pallarenda, AIMS's original site.

To his great credit, Max Day visited AIMS from time to time to see how it was getting on, and even after his hundredth birthday in 2015, he remained interested in the people he knew from those early times. I used to take him around the labs on these visits and so heard much about the aspirations of the institute's founders, some of which seem rather quaint in retrospect: the top scientists would receive a higher salary than the director; no staff would be permitted to hold grants from external sources; the primary commitment of the institute would be to providing the science for the future management and conservation of the Great Barrier Reef. The first of these never happened, the second lasted three years, but the last was reality for about twenty.

Money was virtually unlimited during those years, something I found intimidating because it meant that there was no excuse for any scientist not to achieve their wildest dreams. Nevertheless, it wasn't all rosy at the beginning because of an ongoing war between the scientists and the administration, with the council occasionally butting in with a left-fielder. Several of the original scientists were Americans who loved the Australian way of doing things, except when it came to being told no by a bureaucrat. They stayed for a few years, got sick of it and went home. I was just glad to have my job, and soon learned how the games were played.

'Hey Charlie,' said George Melville one afternoon after a council meeting, 'listen to this.'

I knew he had been recording council meetings, supposedly to help him draft the minutes. He took his tape recorder from his desk drawer and knew precisely the right place to go to.

'I'm concerned,' I heard the chairman say, 'that a few of the scientists seem to be just doing ordinary work. What's Veron doing? We do cutting-edge science here, not his sort of thing. I think his work should be reviewed.'

Six months later I was walking past George's office again when he called me in. Out came the tape recorder.

'We need two or three more scientists on staff like Veron,' the chairman said, 'people who are getting recognised. We should give him a promotion to set an example. Make a note of that, George.'

I didn't exactly have a meteoric rise at AIMS. In fact I stayed on the bottom rung, the lowliest of the scientists, longer than any other scientist in the institute's history. No matter, I had a good job if I could put up with George – and I found that easier than most. Then I went from the bottom rung to the top in eight years, which is also a record.

Despite the good intentions of the founders of AIMS, some bad ideas followed. One was that scientists should be grouped into programs and not treated as individuals. That was not good in my view because it tended to demean the individuals concerned, but more importantly, I always copped the job of leading a reef research group, and that covered about half the institute's scientists. At first, being a program leader meant little more than signing leave application forms, a problem I solved by signing a pile of them and leaving them on the reception desk in the institute's lobby for anybody to fill in. The next job wasn't so easy; I had to approve (what a repugnant word that is) applications for promotion. My natural inclination was to just approve them all, especially if they fitted somebody else's guidelines, but this didn't go down at all well with the administration, who wanted promotions to be their prerogative anyway. So, drawing a little inspiration from the CSIRO, I typed up a personal performance evaluation form, which staff needed to fill out before being considered for promotion.

I hadn't foreshadowed where this would go in the hands of the administration, for within a couple of years a performance evaluation was a compulsory annual event for everybody, which meant that everybody had to have a 'supervisor'. Then everybody had to have a 'next-level supervisor'. All very hierarchical. It meant that a supervisor could report on everybody they supervised, and vice versa. This quickly became the weapon of choice for staff to bludgeon each other.

Still, life at AIMS was seldom without interesting interludes. One morning, Robyn Williams, presenter of the popular radio program *The Science Show*, walked into my office. He'd intended to interview me about my work but now had to hurry to catch a plane; the interview would have to wait. But in the half-hour he had, would I talk about something – anything – interesting? He placed his tape recorder on my desk and switched it on.

I talked about something I did indeed find interesting: how people had once lived under the reefs of the Great Barrier Reef. It was, I thought, a fascinating subject, for when Aboriginal people first came to Australia the last ice age was in full swing and the sea level was far below where it is today. The whole of the Great Barrier Reef was high and dry and people would have hunted on it for thousands of years, and presumably used the limestone caves under the reefs for shelter.

I was delighted when my talk was broadcast on *The Science Show* a few weeks later. AIMS's newly appointed director (this was in the days after Red) was not. He had just returned from Canberra, where he had been talking up the institute to politicians, and when he heard the broadcast he assumed that I'd spun some wild, April-fool-type yarn. I was summoned to his office.

The director looked up at me and just started swearing. Something about me making AIMS the laughing stock of the nation. When I shrugged that off he turned scarlet and demanded I clear my desk and get the hell out of the place.

Sacked again? This was getting monotonous.

But when other members of the council were asked for their agreement to sack me, one of them said he thought my talk was 'really interesting', and that it was good to see an AIMS scientist 'thinking outside the envelope'. This guy was a geologist and would have known about sea level changes, if not the history of

Aboriginal people. That afternoon a piece of the director's notepaper arrived in my pigeonhole with 'Sorry Charlie' scribbled on it.

My file remained with George, who presumably was enjoying the whole affair, until the next council meeting. Shortly after that I was promoted, so I sent the director a note saying how good it was to see that AIMS had luminaries on the council who knew something about science . . .

But it all went downhill again a few weeks later when the governor-general, Sir Zelman Cowen, visited AIMS. The director and his status-conscious wife were in a great tizz over this – we were all told to dress neatly and then casually stand at strategic places around the institute as the official party made their tour. When the G-G and his wife came to my lab I was gleefully waiting. He stopped, looked at me and said, 'Charlie Veron – you work here?'

'Sure do, Big Z,' I said.

Sir Zelman had been vice-chancellor of the University of New England when I was there. He'd had possums in his roof, which Ric and I were asked to catch. Later I got to know his family a little. Big Z was our nickname for him, which I think he liked.

I left the director and his wife, both aghast at my insolence, guessing about that one.

The Reef expeditions

My field trips up and down the Great Barrier Reef blur together because I did so many between 1973 and 1980 and they reached into every corner. In those years money flowed freely, allowing me to study The Reef in its entirety. I'm probably the only person who's ever had such opportunities.

Long trips, scheduled at the right time of year to places I particularly needed to study, and with the right team of helpers aboard, took a lot of planning, and couldn't tolerate delays because of the timing of the monsoons. I also had the relentless demands of the monographs. I never had to cancel a trip because of illness or injury, but a couple of times that prospect came too close for comfort.

One day, three weeks before my fifth trip to the far northern Great Barrier Reef, I swerved to miss a wallaby while riding my motorbike home from work and came off at high speed. The bike was a wreck, one side of my helmet was sheared off, and my left collarbone was poking through my blood-soaked jacket. A colleague took me to hospital, where I was strapped up but not, I thought, the way I should have been. So I went to a bone specialist, who did a better job and told me to keep my collarbone strapped for two months.

'I can't,' I said. 'I'm off on a month-long diving trip and we're leaving in three weeks.'

'Mate, you're not going diving anywhere in three weeks. Forget it.'

'Charlie, you're not going diving,' said the personnel officer at AIMS, clearly thinking of liabilities, a fledgling concept at that time. But the diving officer, an adventurous type who understood the scientific value of the trip, sided with me, no doubt because he was me.

It was a good trip, made possible because I had an old 'long johns' wetsuit that could be hitched over my undamaged shoulder while leaving the other shoulder free. There was no problem getting into the water – I just jumped – but getting out was another matter. I managed to slowly pull myself up on the leg of an outboard motor and then climb painfully into a zodiac.

The bone specialist occasionally treated other people from AIMS, always asking if they knew the idiot who went diving with a smashed collarbone. I admit this particular bone remains a little unusual, especially as I broke it again years later while showing my daughter how to do handstands, but the collection from that trip, with a dozen new species, sure made it worthwhile.

Another diving trip was memorable not for being painful, just dangerous. It was to the southern Great Barrier Reef, so different from the north it feels like it's in a faraway country. The Reef is much wider and the water deeper. Cays are strewn throughout the central region and high, or continental, islands are scattered along the coast. I have only been to the outer reefs of the far south a couple of times, and this particular trip was the maiden voyage of the *Hero*, the homemade steel boat of Cocky Watkins, a seaman of a bygone era when there was a living to be made from plundering wrecks and catching whatever happened to be around.

I wanted to go to the Pompey Complex, 200 kilometres from the coast. This has the biggest tidal range of eastern Australia, which, given the mass of reefs the tide surges through, makes it one of the most dangerous places for boats imaginable. We arrived in the dead of night, in a howling gale and thundering rain. Most of our company were seasick beyond caring.

'Fuck this place – haven't got a bloody clue where we are!' Cocky yelled at me over the noise of the wind and waves as his boat spiralled around in a network of currents we both knew could reach 8 knots, and which made his echo-sounder jump from 90 metres to near zero in a second.

There being no satellite navigation in those days, just the same old scarily inaccurate charts, we anchored taking pot luck, and Cocky went to bed muttering that there was nothing else he could do. I had no idea what else might be done either, but I spent the night in the wheelhouse, in the dark except for the ghostly glow of the boat's console lights. Outside it was pitch-black and the noise of the gale thrashing the wheelhouse drowned out even the monotonous chugging of the boat's generator. The compass kept swinging wildly and the boat kept lurching this way and that. Dawn brought a scary sight, for by then the tide was out and we were almost enclosed by reefs. These formed such a barrier – 100 kilometres long – to the ebbing tide that the sea level was visibly higher to the west than the east.

The Pompey Complex remains imprinted on the minds of all who venture there: dangerous and exciting, with caves, caverns and cliff faces amid boiling currents twisting and turning down a network of deltaic channels that thread their way through masses of reefs of every shape and kind. All very fascinating and with a geological history that has yet to be told.

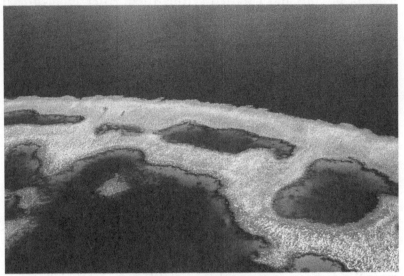

A wall of reef, Pompey Complex.

Reefs like the Pompey Complex aside, the more diving I did, the more carefree I was, and ultimately I just watched out for sea snakes. Collecting corals often means hammering away with a chisel, and that inevitably stirs up a cloud of sediment and makes a great deal of noise. If a snake should come along to investigate – and they tend to be curious – it would likely be whacked in the face by my fins. Such treatment will make any snake angry, and unbeknown to me they often started attacking me. I once saw a photo of myself hammering away oblivious to four furious snakes on me. They can't bite through a wetsuit, but even so, they are horribly venomous, much more so than land snakes.

Once, on a trip to the Swain Reefs of the southern Great Barrier Reef, Len and I were swimming over a wide expanse of sand about 15 metres down when we both stopped to watch a snake gracefully swim up to the surface for a breath. When it started down it saw my bubbles and headed straight for me. It's easy to fin snakes off, so I began swimming gradually upwards in a wide spiral, but the snake kept trying to get at me through the turbulence of my fins and I started getting tired. After about fifteen minutes of this it saw Len's bubbles, left me and went straight for him. Len did as I'd done, until, on the surface and exhausted, he grabbed the snake and flung it into our boat, where it could wait until we chucked it back. I sometimes read in books that sea snakes are not aggressive: that's mostly true, but not for divers working underwater and creating a lot of disturbance, especially in snakes' mating season.

As for sharks, they may well be dangerous in some places, but in all my time diving on reefs, keeping close company with thousands of them, including tigers, bull sharks and hammerheads, I've only been convincingly attacked once. We were at Tijou Reef again, doing a transect of the reef flat using a long measuring tape. We'd chosen a place where, according to our aerial photograph, there was a large lagoon within the reef flat. Our job could be done on snorkel as it was all shallow, except for the lagoon.

We reached the lagoon late in the afternoon, and as we had a scuba tank with us, I decided to go down the edge to take a look. It was surprisingly deep. I reached a depth of about 15 metres when two sharks turned up. They weren't particularly big, 3 metres or so, and would have been of no interest except that they started working themselves into a frenzy, zipping around at terrific speed, arching their backs and making sudden turns. They were clearly thinking about having me for dinner. I was able to fend them off with a heavy chunk of dead coral and by backing myself into a crevice in the reef.

When I reached the top of the reef, having been protected all the way by the crevice, the sharks attacked repeatedly, one crashing into me, open-mouthed, with such force that I was winded. The water surface was about 3 metres above the reef, so picking my time as best I could, I swam upwards as fast as I dared – going up fast being potentially lethal for divers – but when I surfaced there was no boat. I yelled out, then fled back to my crevice.

I hadn't seen the boat, a tinny, because it was between me and the setting sun, but fortunately my buddies had been watching the sharks. Almost immediately the boat appeared above me, and again I surfaced as fast as I dared, and vaulted over the tinny's side with my weight belt and tank on, a feat I have never remotely been able to repeat. My friends had been thinking that being chased by sharks was a joke – we'd been diving with them for weeks – until I pulled off my wetsuit. Parts of my chest were already turning blue.

The following morning, five of us, all on scuba, me at the back, returned to the edge of the lagoon to take another look. Sharks came barrelling in. I took one blurred photo before clearing out – it had seventeen grey reef sharks in it.

Years later I overheard a conversation at lunch about the leg of an outboard being attacked by sharks. Where did that happen? Tijou Reef lagoon. There's something weird about that place: a shark nightclub, perhaps?

By 1977, after working up and down the Great Barrier Reef for four years, I had developed a yearning to go further afield, especially south to see how coral diversity dropped off at higher latitudes. Lord Howe Island, about the same latitude as the Solitary Islands but 600 kilometres from the coast, has the southernmost coral reef in the world. I also knew that it was considered to be one of the most beautiful places in all Australia.

'Where's that?' Red asked when I sought his approval to mount an expedition there. My job was to work on the Great Barrier Reef, not elsewhere.

'Right at the southern end of the Great Barrier Reef, more or less,' I said, aware that Red's Australian geography wasn't the best.

In January the island's schoolroom would be empty, so I arranged to borrow it. This way our team, plus Kirsty, Noni and a friend of Noni's, had a place to stay for nothing. We spent a glorious month on the island, diving and working on the corals from a zodiac we brought over. I'd been spending too much time travelling and writing monographs; this was a trip to be enjoyed. And we did. The divers spent most of the day out on the water, then we'd all go for long walks, watching tropicbirds, terns and shearwaters doing their last fishing for the day, and twice seeing the little flightless woodhen endemic to the island, and at one time facing extinction. The island, listed as World Heritage only a year after the Great Barrier Reef, was like no other I'd ever seen, hosting a unique, subtropical mountain forest whose plants were akin to New Caledonia's as much as to Australia's, and with a geological history to match.

Not surprisingly, the coral communities of the island were also unlike any others. Except for a single species only found in temperate waters, all were immigrants from the Great Barrier Reef, as were the corals of the Solitary Islands, but the species mix was very different. I believe the corals probably came and went from the island at frequent intervals, geologically speaking, coming in on the warm, south-flowing East Australian Current from the Great Barrier Reef, and being wiped out by the cold, east-flowing Antarctic Circumpolar Current from southern Australia, should that make an incursion north of its usual path. This would explain why several species only had one or a few colonies on Lord Howe, not enough to be a self-sustaining population.

We found colonies of one such rarity, each brightly coloured with individual patterns. I photographed them all but collected only one specimen, fearing that the species might be on the verge of extinction. I called it *Acanthastrea lordhowensis*, not the best of names, because I later found it in many other countries, and in some it was common. Being so beautifully colourful and easy to culture, it has become one of the most prized species for reef aquaria, with thousands of colonies now to be found throughout Europe and America.

My month on Lord Howe also led to a discovery of a different sort. A week after I arrived, the local paper (a page or two of foolscap printed with a Gestetner machine) included an article about our team. Shortly afterwards, I was filling scuba tanks on the beach when a middle-aged man approached. I'd just watched him and his buddies row ashore after a day's fishing. 'Are you Dr Veron?' he asked. And when I said yes, 'Any relationship to Donald Veron?'

'He's my father.'

'You know, everything your father says is true – the whole lot. But I doubt it will get him anywhere,' he added. He turned to walk back down the beach.

'And who are you?' I asked, somewhat taken aback.

'I'm Justice Kirby, chairman of the Australian Law Reform Commission.'

Perhaps Dad's not nuts after all. He'll be interested in this.

But he shrugged it off when I told him. My father knew what he was doing and why. Along with many others, he went on to play a part in driving the then police commissioner, Frederick Hanson, out of his job, revealing the corruption that brought down the Askin government. Hanson later suicided. I still have a tea chest full of documents from this long and bitter battle; they make quite a story.

The following year, 1978, AIMS commissioned its first oceangoing research vessel, which was to be named after Sir Henry Basten, the first AIMS chairman. However, Sir Henry declined the honour and so it was called the *Lady Basten*, after his wife, a name more suited to a ferry than a research vessel. No matter, I was asked to be in charge of its shakedown cruise, one that needed to go well because there'd be a maritime inspector aboard who had to approve all operations involving safety at sea. With a free hand to take her where I wanted, I chose a zigzag track across the Coral Sea to the Chesterfield Islands, near New Caledonia. We would be at sea for about a month.

This turned out to be the sort of trip I love. There are not many reefs or islands in the Coral Sea but each is very different from the others, and we never knew what to expect. Most of this sea is now a marine reserve, nearly a million square kilometres in area, but when we went there almost nothing was known about it. It was fabulous wilderness; there were no other boats, just one extraordinary place after another: thickly forested cays, grassy cays like Raine Island, and sometimes just reefs without cays, each unique and swarming with life. I'd love to have had the time to make a study of the corals of each island and reef, something I still need to do.

The Chesterfields proved to be a great discovery, the most pristine reefs I'd ever seen and ever will see. They were swarming with fish of every size and description, so much so that I had to constantly shoo them away when photographing corals. On one very memorable occasion I dived from the back of the *Lady Basten*, did my work, and was using the last of my air to collect a species of algae a colleague wanted when I heard the thump of a grouper, a common warning sound they make with their tail. I looked up to see the back of a Queensland grouper slowly rise from behind a nearby knoll of reef. It turned and came slowly towards me.

I had never heard of a grouper big enough to swallow a human, but this one had a head almost the width of a whale shark's and a body the size of a car. I would have scrambled for cover in the coral if I'd had any air left, but as I hadn't I took off my tank, thinking it might like to swallow that first. I had seen that groupers catch their dinner by suddenly opening their mouth and sucking their prey in, and this one could clearly swallow both me and my tank in a single gulp. The fish ignored the tank and moved right up against me, at one stage its huge eye almost touching my mask. I didn't have far to go before reaching the safety of the *Basten*, but just then that felt like a very long journey. When I reached the surface the grouper was still right beside me, but then it turned and swam slowly back down. Maybe it had just wanted me out of its territory? I was only too happy to oblige.

All this had been witnessed by several people on the *Basten*, and a member of the crew was able to get his camera in time to take a blurred shot; the grouper was well over twice as long as me, more than 3 metres. I know of no record of such a fish, although there have been unconfirmed reports of exceptionally big Queensland groupers.

Of course a lot of people think this is just my Fish That Got Away story.

Inge of Orpheus

A decade ago, James Cook University's Orpheus Island Research Station was not as upmarket as the Heron or Lizard island stations, but now it's a place to be proud of. I'm proud of it too. Here's why:

I first had a good look at Orpheus from the wheelhouse of the *James Kirby* in 1973, on the first day of the Stoddart expedition. We reached the southern end of the Palm Islands about six hours out of Townsville, with most of our company so seasick they were already thinking the expedition wasn't such a good idea. So I stayed in the wheelhouse with Davie Duncan while he told me about the history of each of the islands we passed. The main island, Great Palm, was an Aboriginal reserve with a horrific history of violence and alcohol abuse. Another island had a leper colony run by an ancient Canadian nun – I'd had no idea Australia still had leprosy.

Davie had been fishing around the Palms for most of his life and knew every detail of all the coastlines. So, just to make a point, he brought the *Kirby* close to the western side of Orpheus Island, one of the northernmost of the Palm group. With a rocky peninsula straight ahead I wondered what he was up to, until we came to within a stone's throw of the rocks and the beautiful white beach of Orpheus's Pioneer Bay opened abeam of us.

'They reckon a mad red-headed Austrian woman lives there,' said Davie. 'Hangs the red light out for fishermen, so they say.'

I swept Davie's binoculars along the lonely beach but saw no naked nymph frolicking in the waves. However, it did occur to me that this place might make a good base for diving trips. Davie thought so too, but then we both forgot about it.

The following year I took two boatloads of students there. We camped on another island and went diving on the west coast of Orpheus from a tinny and an inflatable, both old wrecks. We'd arranged to meet in the middle of the island for lunch, but a storm came barrelling in. I was in the old inflatable with two others, battling to keep it going in thundering rain, and spent an anxious hour searching for the tinny. Then, through a break in the downpour, we spied it high on the beach of Pioneer Bay.

We went ashore. There were no students we could see, only the remains of an overturned dory, and a sandy path disappearing into the scrub. Up the path we went.

'Hello,' came a greeting in a cheerful, heavily accented female voice. 'I suppose you vant der crust also? How many crusts can one loaf have? Can you tell me dat?'

It was then I remembered Davie's comment of the year before.

'Davie Duncan reckons some woman lives here,' I whispered to Susie, one of the students.

'Who?' asked Susie.

'Don't know. He just said she's an Austrian – some sort of hermit. Well, if she's friendly, at least we'll get out of the rain. I'm freezing.'

The woman in question emerged from the door of a small house that looked as if it belonged in a Hans Christian Andersen fairytale. Maybe a witch's house?

'You are very, very vet,' she observed. 'I sink you can't get any vetter even if you stand in der rain all day. Take your clothes off. I have plenty of honey und bread but not de tiniest crust is left. Vy do you not tell me you vere coming? Den I bake lots of bread viz lots of crusts! You are velcome even if I don't know you are coming. But dis bread I have just dis morning baked.

'I know about you all,' she went on, naming the three of us. 'I am Inge, I get you towels, you look like drowned rats.'

I struggled out of my wetsuit and my two companions did the same.

'You live here all alone?' I asked when Inge returned with some plates, a couple of towels, and a platter piled with rather mutilated loaves of bread. She was of medium height with a serene, gentle face that radiated goodwill and empathy.

'Never! I am alvays Yellow und Mausie viz.' She nodded towards two dogs. Yellow looked like a cross between a labrador and a dingo. He was obviously named after the colour of his very expressive eyes. If you don't try to pat me, they said, I won't bite you. We found instant understanding.

So began my long friendship with Inge Moessler, one that was to become very important to us both. She wasn't mad, nor did she hang out the red light, instead she was one of the most interesting people I have ever known. I came away from that first trip with my head spinning. Being a scientist with an uncompromising approach to anything supernatural, I had a hard time believing she was as clairvoyant as she appeared to be. No doubt it was living alone on an island that helped make her different from anybody else I'd ever met.

It didn't take me long to realise we had discovered a perfect base for field work. There was a level patch of ground where we could pitch a permanent tent, and an old shed where we could safely keep our scuba tanks, compressor, and other diving equipment.

Over the following years, I returned to Pioneer Bay most months and so came to know Inge very well. After I was through with the day's diving, she and I would usually sit on the overturned dory on the beach and sip wine as we watched the sun go down. Inge always did this; she said it was her way of going to church, a special place where she could feel particularly connected with Nature. I could see why. Between daylight and night, the beach, the headlands, the sea, the sky, and the distant sawtooth row of mountains of the mainland all change colour in unison so rapidly that it's hard to absorb one panorama before it turns into another. It's of no comfort to reassure yourself you'll watch it again tomorrow, for tomorrow it will all be completely different. Sometimes the panorama is overwhelmingly red, the sea and sky the colour of blood and the mountains black. The next night, maybe there'll be a silver sunset, with cold clouds almost devoid of colour. For me, most beautiful of all is the sunset through heavy, low, water-laden monsoon clouds. These go deep purple, higher clouds go pale pink, both are pierced by shafts of bright yellow light. The sea, in response, is silver, perhaps slashed with red while the mountains stay black. All a riotous display of nature gone berserk, as if painted by a heavenly maniac intent on proving that his genius is indeed infinite.

At these times I learned a lot about Inge's incredibly convoluted life, which was packed with highs and lows few could imagine: being banished from her aristocratic home when she gave birth to a daughter while still at school; surviving the horrors of wartime

Germany; having an amalgam of husbands and lovers in Africa, Brazil and Europe; living under sheets of corrugated iron next to a Sydney rubbish dump; moving to Alice Springs with a roo shooter, before coming to Orpheus Island to find solitude.

After the first glass of wine Inge would start chatting about this and that. Her accent made her stories hard to follow at times, especially as most would jump from one continent to another and get entangled with other stories. After a second glass, she'd often include snippets about her love affairs, of which she sometimes gave me intimate accounts. If she downed a third glass, she would usually have forgotten by morning that she'd told me anything.

I discovered we had the same love of classical music, particularly lieder, and more importantly the same love of Nature. In her own way she was somewhat religious, with a personal deity much like the god of my childhood. It was intriguing to be with someone who communed with Nature as I had done all my life, and who used to effortlessly meditate in her younger days as I occasionally still did. Yet she was the very opposite of me when it came to thinking about the natural world. I always yearned to understand it, whereas she was content just to feel it. She had a bedroom with two walls missing so she could be part of the surrounding bush. A pair of small birds had a nest above her bed and had been coming back to the same nest for years: I knew they were sun birds and also a little of their nesting behaviour; Inge just loved being accepted by them. She always knew where they were and what they were doing. Her little python came and went as he pleased, as did dozens of other animals, including Lisle her hen, which laid an egg somewhere in her bedroom most days.

As time went on, Inge came to know me as well as any person ever had, or at least she would know what I thought and why I thought it. She was the same age as my mother would have been, so in some vague sense we had a mother–son relationship, which helped me through the nightmare that lay ahead.

By 1978 I had used Inge's place as a base for diving dozens of times. I never thought of it as anything other than Inge's place, which was foolish in hindsight as I should have realised that only a very rich person could own such a place. And so I was shocked when she told me that her lease, which was an oyster lease, had expired and that she had been ordered to leave the island within a year, removing her house and all other property when she did. I wrote to government departments of all sorts to try to get this changed, but without success – Inge was going to have to leave her island.

Then one evening at sunset she murmured, 'Dis is such a perfect place for students,' and right then I decided to try to convince James Cook that it needed a marine station at Pioneer Bay.

Within a month I had Cyril Burdon-Jones (no less) in a tinny (of all things) heading from the mainland to the island. After running aground several times, we finally got ashore and received a warm welcome from Inge, who then castigated me for not bothering to check a tide table; we had come when a spring tide was at its lowest. She had lunch ready for us, laid out on her wobbly trestle table under a big rainforest tree that grew in front of her house. The professor made a tour of inspection.

'The buildings,' he said, 'are not of a nature one normally associates with universities. And the generator,' he observed, 'appears to be in need of the attention of an electrician, and perhaps of a mechanic also.' Of the 'toot' – Inge's gorgeous, falling-to-bits outdoor dunny that afforded a commanding view of the surrounding rainforest – he offered no opinion at all.

Inspection finished, we started lunch. Lisle the hen didn't help by insisting on pacing up and down the trestle table where lobsters and coral trout were set out, but all things considered, the meal went off with fewer mishaps than normal for Pioneer Bay.

Once back at the university the professor announced, not unexpectedly, that he didn't think Pioneer Bay would make a suitable site for a marine station. 'There are no proper facilities as would be required by a university. But,' he added, 'I did enjoy the trip, especially meeting Inge. She's a very remarkable person.'

Sarcastic old goat.

But I was wrong: the professor and Inge had actually hit it off, as I was to see.

In the meantime I told Ken Back about Pioneer Bay. A keen yachtsman, he sailed over to check it out and not surprisingly fell in love with it. He realised it would make an excellent marine station for the university and immediately set about getting a new, long-term lease for Inge's land. I promised to help on the condition that Inge would be allowed to live there.

And *that's* how the Orpheus Island Research Station came into being.

Inge was given the unpaid job of manager of the station. Poor university – the administrator found her hard to take, so much so that I had to agree to be on a management committee so that the university could manage her. Me on a management committee! I had long ceased working for the university by then, but we managed well enough, until in 1986 her agreed tenure on the island expired, and this time she really did have to leave. The administrator rejoiced, the students protested, and Inge thought her life had ended.

John Coll, a professor of chemistry at the university, and I bought Inge a little house on the bank of the Herbert River, just across the sea from her island. It was a rather ramshackle place on wooden stumps, but full of character and with a nice garden, which she could enjoy with her dogs. Of course she wasn't happy there, it was nothing like her island.

A couple of years on I was taking some students to the research station, which by then was well established, and stopped to see Inge on the way. It was her seventy-fifth birthday and we'd planned a celebration. She was in particularly good form, telling the students one mesmerising story after another, all in her heavy

Austrian accent, difficult for those who didn't know her as I did. Her stories were hard for the students to keep track of, especially as I frequently interjected with facetious comments, having heard the unabridged version of whatever she was on about many times before.

'Charlie,' one of the students blurted out, 'why don't you write all this down so we can follow it? Inge's lived more lives than an alley cat.'

'Good idea,' I said, 'I'll write a book and call it *Inge the Alley Cat*. Eh, Inge?'

At such times, and there were plenty of them, Inge would just look at me with her calm grey eyes. You'll keep, they would say.

Next morning, while we were doing the washing-up, sure enough, she rounded on me, a big smile on her face. 'So you are going to vite a book about me, are you?' she said in her haughtiest voice. 'And just vot vood *you* know about *me*?'

Over the years I had, as is my nature, pieced Inge's life story together, and it sure was an interesting one. Thus armed, I gave her the whole lot without mercy, including the juicy bits – especially the juicy bits. Inge just stared at me goggle-eyed. 'I never told *anyone* dose tings!' she stammered.

'Didn't I tell you I'm clairvoyant?' said I. 'But don't worry, I can make up bits of fiction about the parts you're really ashamed of.'

'I am not ashamed of my life!' she shouted at me, banging a frying pan down on the kitchen bench.

'That's good,' I said, continuing my game, still without a hint of mercy, 'it should make interesting reading.'

I had no intention of writing a book about Inge or anyone else, but I wanted to learn to touch-type and so it more or less started to write itself. After about a year, with much joy in anticipation, I printed it out for Inge to read, but disappointingly she didn't explode; she just calmly corrected points of detail.

A month later it all backfired. Inge's daughter Ingrid, the child

Inge had while still at school and whom she hadn't seen in thirty years, came to visit her from France. Inge promptly sent Ingrid to Rivendell to see me and to read 'her' book. I told Ingrid it was only something I'd written for fun, and left her on the terrace to get on with it. A couple of hours later I went to see how she was doing and found her with tears streaming down her face.

'Charlie, if you publish this it will break my father's heart,' Ingrid sobbed.

The kindly man she had known as her father was Inge's first husband, a paediatrician, by then very old and living where Inge had left him, in Ethiopia. Ingrid had just discovered that her biological father, in stark contrast, had been a Nazi war criminal of the worst kind. I was shocked at having caused so much grief.

Many years later, Inge phoned me to say that her husband had died. 'So now you can finish my book,' she said.

I pointed out that it had only existed on a thing called a floppy disk, which computers were no longer able to read.

'No matter,' she said cheerfully, 'I still have dat copy you gave me – a little eaten by silverfish, but never mind. And I've fixed more of your mistakes. You never listen to me,' she added.

Inge had transformed her little house into a place she cared about. It was stuffed with knick-knacks, had vines on the walls and pot plants everywhere, big green frogs in the toilet, and all manner of birds coming and going through her ever-open windows: it was a place that Robinson Crusoe would have appreciated and she had come to call home. But her health had steadily deteriorated, her spine was falling apart. After many battles I installed her in a nursing home in Townsville, despite her insistence that she would kill herself by starvation. I saw her every day; she always wanted me to read her some more of her book, and she usually had me correct something or other. She was convinced that it would be made into an epic movie starring Meryl Streep. However, as threatened, she refused to eat anything.

On 10 January 2003 I read her the last few pages of her book.

'I sink you are right,' she murmured.

'Right about what?'

'About God. Dere is no God.'

She fell asleep. Late that night the nursing home rang to tell me she had passed on.

'Thank heaven for that,' was all I could say. And I meant it. Watching someone suffer as much as she had was gut-wrenching.

Inge and I were so different in many ways yet so very much the same in others. I could talk to her without words. We could listen to lieder or opera as if we were one and the same person, saying nothing because the music said it all for us both. She alone had the know-how to stop me being overtaken by the frenetic world of the day and reconnect me with Nature, reconnect me with who I really was.

I still have her visitor's book from Pioneer Bay. It's crammed with jokes and anecdotes in the many languages she spoke. It also has words from grateful people she rescued from themselves, allowing them to find a new life and be free.

Inge was published just after she died.[13] Ken Back, John Coll, Terry Done and Issie Bennett added words about it on the back cover, but somehow I doubt that a copy ever reached Meryl Streep.

We spread Inge's ashes on the beach at Pioneer Bay. I then sat on the side of a dinghy and watched the water ripple over the coral. The tide came in and went out but I didn't notice.

Two daughters

By 1977 Rivendell had been transformed from a new house into a comfortable home, its solitude and appearance evidence of a way of life utterly apart from town. By then we had electricity, so the kitchen and bathroom taps worked, courtesy of a pump that drew water from the river. We had cleared a lot of chinee apple and the grounds looked more like a secluded park than a patch of scrub. When we weren't working we read books, talked or, best of all, listened to Noni getting on top of yet another piano piece.

My own little patch of seclusion by the river continued to be very important to me. It fell into deep shadow before sunset, well ahead of the surrounding property, allowing the water to sparkle and the treetops to shine. Birds of all descriptions came from everywhere and the river's surface constantly rippled from fish and turtles feeding, and the occasional snake daring to make a crossing. There were no foreign noises. After a day's work, problems seemed to get left behind as soon as I entered and sat on my tree root, or perhaps a folding chair. I could sit there, soak it all up and let my mind drift into emptiness. Just for a while. Then I would resurface, calm and relaxed. I might then think about a trip or perhaps plan an article.

In all, Rivendell was living up to Tolkien's description. Our former life in Townsville seemed a long way off; we felt we had started anew and so we decided to try, once again, for another child. This time Kirsty's pregnancy went well, but it was not to be. The miscarriage came at short notice, so she wasn't battered around like the first time. It was a setback for sure, but with the life we were leading she recovered quickly, mentally and physically. Noni, on the other hand, clearly felt it was her own personal loss.

When Noni was eight I arranged to do some diving for a film crew who were making a documentary about reef research. The crew, Noni and I went out on the *Lady Basten* to Keeper Reef, about 75 kilometres from Townsville. Noni was very keen to try scuba diving and although I'd never heard of anybody so young using scuba I couldn't see why she shouldn't give it a go – she was a good swimmer and had a steady nerve. I gave her a crash diving course on the back deck, showing her how to clear her ears, how to get water out of her mask, how her regulator worked, and explaining gas pressure – why we should neither stay down too long nor come up too fast. She had a wonderful time; her only regrets were that there wasn't time for a second dive and she didn't see a shark.

Noni did more diving at Orpheus Island. This time I borrowed a small tank for her to use, and from Inge's boat we could dive down the edge of the reef, well beyond the beach. Noni was always keen on collecting things and would point out bits of coral for me. We went on several dives this way, and then she gave me the fright of my life – she vanished. For the first time ever I found myself fighting panic. She had no snorkel or life vest, we were near deep murky water and she would not know the way to go. She was totally dependent on me looking after her.

I went for the surface; she wasn't there. Battling to keep my head in some sort of working order I started to swim in what I hoped was a spiral out from where I'd last seen her. After what seemed an eternity, there she was, watching some fish and waiting for me to catch up. Relief flooded every part of my being. I had to clear my mask, not of seawater, but of tears. On the next dive we were connected by a float rope.

She could have drowned and it would have been my fault. I wouldn't have been able to live with that.

Noni had so many pets around Rivendell it was starting to look like a zoo, and thus I was a little surprised when they suddenly took second place after she went horseriding one afternoon. She immediately wanted her own horse, which we couldn't afford, but she went to a riding school and for a while riding even started to compete with music in her extraordinarily busy life. No mount seemed too much for her and she especially liked jumping. Kirsty, having been brought up with horses, thought this was a great idea, although Noni managed to terrify me when her horse, at full gallop, veered suddenly to jump a fence. She lost control and came off. Undaunted, she chased after the horse.

Fortunately, music won out over horses, but over the previous year or so Noni had developed another passion, which she would not put to rest; she desperately wanted a brother or sister. She kept on about it and eventually got her way. Katrina was born on 25 March 1978. Noni was beside herself, but our joy at finally having a second child was immediately mixed with worry. Katie had respiratory problems, a cleft soft palate, and, though we didn't know it at the time, a hole in the heart.

At first, Katie's problems didn't seem so great. Her paediatrician convinced us that the cleft in her palate could soon be repaired, but she needed to remain in hospital for a while. Like Ruari, she was kept in a humidicrib and given oxygen. But Ruari's troubles and hers were quite different and it was quickly clear that Katie's main problem was a narrowing of the airways due to a developmental problem in the growth of her lower jaw. Her cleft palate meant that she was unable to take milk from the breast, and then it proved impossible to feed her from a bottle, as any milk in her mouth could have got into her lungs. So we had to resort to feeding her with a tube down her nose. This worked well enough but caused all sorts of other problems, not least the need to replace the tube every day, which required a daily trip to hospital until Kirsty learned to do it herself. The procedure was a painful and horrible thing to inflict on any baby, let alone your own, and it often left Kirsty drained and trembling with emotion. More long-term problems followed, one after another. Katie was in and out of hospital and making little progress. She had to battle dreadfully for breath, had repeated aspiration pneumonia, and her heart was failing.

When her paediatrician in Townsville had done all he could, Kirsty and a nurse took Katie to Melbourne's Royal Children's Hospital in a humidicrib. There, for month after month, doctors tried one treatment after another, sometimes with temporary effect, more often with none. Kirsty's life was unremitting torture.

When Noni and I went to Melbourne for a visit I was horrified by what I saw. Although Katie had grown in length, she was by then less than her birth weight. Kept in an intensive care ward, along with another dying baby, she was nothing other than skin and bone covered with a mass of plastic tubes and gadgets. As a last resort, the committee of doctors whose care she was under decided to perform a tracheostomy so she could breathe through her trachea. The operation was declared a success and its effect was immediate. For the first time in her life Katie could breathe easily and her heart wasn't under stress.

We brought her home. There she could be tube-fed, a drop at a time, every three hours; a little more would cause her to vomit, a little less and she would lose weight. But this was a small problem compared with the tracheostomy. Having no air going to her vocal cords she could not cry if something was wrong, nor could she clear her throat. This meant that if her trachea wasn't cleared for her when needed she would drown, in silence, in her own mucus. When she was awake her trachea needed clearing about every half-hour, then her breathing would be silent until it needed clearing again. When she was asleep she normally lasted several hours. Katie breathed through her trachea for eight long months, during which time she had to be watched day and night. Kirsty and I only slept alternate nights, and I learned to sleep so lightly that I woke immediately if Katie's breathing pattern changed. I paced the floor when not asleep, listening to music and helped by a little marijuana, which kept me alert. No alcohol, that was for sure.

Katie had a continuously high temperature that did not return to normal until she was nearly two years old, coincidentally when we all visited Inge on Orpheus Island at Easter in 1980. On that trip Katie began to smile and take an interest in her surroundings for the first time, and then finally she began to feed without a tube.

Noni, for her part, never showed the slightest sign of resentment of the hardship that had befallen us. And certainly it was hardship for her as well as for Kirsty and me. The three of us had neither holidays nor outings together; Kirsty and I were constantly tired, and with tiredness came intolerance. Gradually the early happy days of Rivendell faded, and this must have been all the more apparent to Noni. We knew this and tried to make it up to her when we could, but when life had become a matter of just surviving the next day, Noni's problems seemed small indeed. She called Katie 'Termite', a nickname that stuck, and if she was disappointed at Katie not being the baby sister she could mother and play with, she didn't show it.

But she did start keeping her worries to herself when once she would have talked about them, and she also started not doing her chores around the house, and skipping her homework. She grumbled about her teacher and sometimes about schoolfriends, a state of affairs that continued on and off for months, partly unappreciated by us until her teacher complained that her marks were nothing like they used to be, and that she had become a classroom 'problem'. Her worst subject, apparently, was maths. I found this astonishing because when we were in Melbourne only six months earlier I'd helped Noni with her maths, and within a week she had not only caught up with her schoolwork, but gone right through the entire year's curriculum. The speed at which she absorbed maths amazed me, and not for the first time I wondered what school did for her.

The three of us had a long talk, not about maths, but about her teacher and school and home. These were not small problems for Noni and they were seriously affecting her. It took some failures and a few painful weeks, but she won her teacher round and started being her normal happy self again.

At the age of nine, now at the top of her class in most subjects, Noni decided to take up the cello. She played in the Townsville Junior Orchestra, in her school orchestra, and occasionally in duets and trios with friends. But most of the time she played the piano at home. She gave a winning performance at a music teacher's concert, and then floored us by announcing that she was going to take up the flute. At that time singing had also come to the fore, especially duets with Kirsty. Once again, Rivendell positively rang with music.

As they had in London, music soirées and the stage provided Kirsty with a much needed escape, and she was in one musical or drama production after another, sometimes to great acclaim as the lead singer or actress. Naturally Noni was very interested in all this, particularly after she was asked to play the second-youngest of the von Trapp children in a production of *The Sound of Music* at Townsville's Civic Theatre.

Nineteen eighty was going to be Noni's year. I had planned another four months' study in England and Europe, and this time Kirsty, Noni and Katie were coming too. Noni could afford the time away from school, with Kirsty and I keeping up her lessons, something that would be increasingly difficult the older she got. We would find somewhere for her to practise piano and cello and she could, as she had long wanted, learn French in France. Kirsty, Noni and Katie had their tickets while I waited for a scholarship. I explained to Noni that without the scholarship I couldn't afford to go, and we'd have to make other plans. Noni didn't agree – she, Katie and her mother had their tickets and were going, no matter what. If I couldn't come it was 'too bad'.

I got my scholarship and we arranged with friends to rent half a house on the edge of Epping Forest, just down the road from a riding stable. The year ahead was looking fantastic.

Noni and Katie, 1980.

Just after we made these plans I went to Hong Kong for a week-long coral workshop. Towards the end of the trip I was having an evening meal and enjoying the conversation when the convenor interrupted. With his hand under my arm as if to support me, he ushered me into another room, followed by a couple of others. Hand still in place, he passed me a phone. I heard Kirsty trying to force herself to talk to me but I could make no sense of it. I froze with fear. Then I heard Alastair's voice telling me something I just could not absorb – that Noni was dead, that she had drowned.

I dropped the phone. My companions started talking to me – *at* me. They had been forewarned. What was all this about Noni? Al's words started sinking in. Then Noni started talking to me, but not on the phone.

It wasn't until much later that the full story penetrated what was left of my paralysed mind. Noni's tenth birthday party had been due later that week, but one of her friends was unable to come, so the two mothers and their daughters met for a pre-birthday picnic at nearby Alligator Creek. As usual the girls were playing close by, where the rocks formed an inviting slippery-slide into the creek. Noni was strong and at home in the water. She may have hit her head, we'll never know. Her friend found her pinned under a rock, the water flowing over her. She tried to pull her free but wasn't strong enough. It was Kirsty who did that unimaginably horrific task.

If time cures all, I am still waiting, at least for the closure that many people seek. Nearly forty years on, my memory of this time is fading, but only the memory of facts; my feelings remain as raw as ever.

After that fateful phone call someone managed to find me a flight to Australia. I arrived at the airport the following morning. I hugged Kirsty until my arms ached. How could life do this to her after all she had been through? I had never remotely felt such intensity of grief for another person in my life.

I saw Noni in her coffin. I kissed her face; it was frozen. This is the worst memory of my life.

Noni was cremated on 29 April 1980, her tenth birthday.

I lived courtesy of tranquillisers and sleeping pills, and so it remained for many years. I'd once loved the dawn at Rivendell, but not then. If I drugged myself enough to get a little sleep, dawn only brought reality, and I resented the kookaburras for trying to tell me that the world was as it used to be.

After a month, I was able to get up, go to work, and seemingly have a normal day – even crack a joke. My external world, being mostly work, was indeed as it had been. My inner self was very different but often strangely contented, for like an automaton I would be talking to Noni, even when I was talking to someone else. I don't know how long that lasted, but I remember only too well the dread that started to descend when I became aware that our conversations were getting repetitive, with fewer and fewer new subjects or arguments or points of view. That was almost like Noni dying a second time for me. A third came when her voice itself started to fade. She doesn't talk to me at all now. If that's a cure, please God, I want the illness back.

We didn't touch Gum Leaf or Noni's bedroom for many years. We wanted to remember, not forget. The place where Noni kept all her ducks and her geese is now Noni's Garden. Our friends gave us plants for the garden and they have grown into mature trees. It is a wild, spiritual place. Her ashes are there and it's also where our dogs lie. At the edge of the garden is a tall umbrella tree, the last present Noni ever gave me, for the last Christmas we had together. It was decorated with coins stuck to the leaves with sticky tape, to show me that 'money grows on trees'.

'So now you can buy me a horse,' she'd said.

*

As we had following the death of my mother, we went ahead with our trip to England, supported by our English friends at the university.

Sometimes, when with people we hadn't previously known, it might have seemed that Noni had never existed, reinforced by my complete inability to talk about her. I worked by day at the Natural History Museum, going there on the underground in the dark and coming home to our house near Epping Forest in the dark. Kirsty spent most of her time helping Katie learn to walk and do the baby things she'd missed out on. We all travelled together a bit, seeing Scotland and other parts of England. I particularly remember hearing some chamber music in Cambridge, at a music workshop Kirsty was attending, and being fascinated by an elderly cleaner who knew more about the music than I could ever hope to. But what I remember most is walking down a London street with Kirsty. It was dark and cold and raining. Once again I was in tears and Kirsty, once again, was trying to prop me up. I remember a flood of self-disgust. After all we had been through, *she* was helping *me*. Again. Why wasn't it the other way around? No doubt it often was, but not as often as it should have been, at least as far as I could recall. Sometimes I tried to stand in Kirsty's shoes, and couldn't.

The best part of our trip was coming home. Rivendell had been well looked after by Terry Done's parents, and my father came up to see us as soon as he could. He had changed completely. He'd forgotten the battle for justice that had so preoccupied him since my mother's death and turned himself into a loving father, helping, caring, doing all he could. He was close to me as never before, as was my sister. Noni had become a major part of Jan's life and I feared she suffered in silence, without much help or understanding. She started becoming embittered with her life.

I hope I became a better brother to her than I had been. Jan tried to be cheerful when we were together, but this only worked with the help of alcohol.

For several years I was Townsville's so-called resource person for parents in extreme distress from the death of a child. Such parents are exceptionally sensitive to what they hear, and they even hear the unspoken thoughts of others. At that time it was easy for me to be a guide, and not make the mistakes that many blunder into. But I eventually gave it up because of the toll it took.

For a very long time, when on a field trip, I was gripped by fear if a radio call was for me. I even feared the sound of any incoming radio message, before the advent of satellite phones allowed casual contact. At least this ease of communication now gives me some peace of mind, but I've never been able to immerse myself as wholly in field work as I once did. Worry for my family is always with me. I fear it will be so to the end of my days.

Going west

With most of the Great Barrier Reef expeditions, as well as journeys to the remotest reefs, islands and atolls of the Coral Sea and beyond completed, I had started writing volume three of *Scleractinia of Eastern Australia* in 1978, and that helped me through the remainder of 1980. As many have noticed, the publishing quality is poor – the pages feel like they're made of cardboard and the volume is even the wrong size. I just wanted it finished, and left the production to somebody else. Field work for the remaining two volumes, mostly done simultaneously on those trips, was also in good shape by this time. AIMS treated me well, something I will always be grateful for. The institute forgot that they'd given me four years to produce the monograph series and I rightly reckoned that they wouldn't notice if it took a little longer. It took eight.

In 1982 the fourth volume was published. For a reason I've forgotten, this started another row with AIMS's administration. Again the matter went to the council, and again I received a nice letter praising my work and telling me I had a promotion, and also, this time, tenure. In my experience it pays to have fights with bureaucrats, at least those that look winnable. The fifth volume was finished soon after. Carden Wallace, a specialist on *Acropora* and a friend, later to be curator of corals at the museum of Tropical Queensland, joined me for that.

In all, I was glad to see the end of the series. I loved the field work and seeing so much of the Great Barrier Reef, but the many months of museum work I had to do were deadening. There always had been a divide between the two, and as long as both museums and reefs exist I'm sure there always will be.

The monographs certainly earned their place in the history of coral taxonomy. The first volume was substantially superseded after further work, but the series remains a foundation reference for most groups of corals, especially the big, complicated genera *Acropora*, *Montipora* and *Porites*, which dominate most coral reefs throughout the Indo-Pacific.

With the monographs finished and my job having survived against the odds, I needed a new direction. The Western Australian coast beckoned, but not just because it was there. The basic question was: did the corals of the west form a mirror image of those of the east? Were they the same species, with the same abundances? Did they occur in the same habitats? And, most interesting of all, did they have the same pattern of latitudinal attenuation? If not, why not? The taxonomy and the data on distribution and ecology we had assembled in the east made a foundation for the big goal I'd long had – to build a unified coral taxonomy for the whole Indo-Pacific. On the Great Barrier Reef we'd made a start for the Pacific; with the corals of Western Australia we would maybe do the same for the Indian Ocean.

I first went to the west in 1981, partly to catch up with Ric How, my old buddy from University of New England days, and his family. I also wanted to go to the Western Australian Museum in Perth, where Ric was a curator, to see the coral collection there and to meet with staff who knew the coastline. Barry Wilson, a former director, then offered to take me in his four-wheel drive to Ningaloo Reef, about 1000 kilometres to the north. We got bogged at most creek crossings but as there weren't many creeks to cross we made it easily enough.

At that time it was generally believed that the big ocean currents that border the three continents of the Southern Hemisphere – Africa, Australia and South America – run north along their west coasts and south along their east coasts. A glance at an old atlas Barry had showed that this was indeed the case for Western Australia, so I wasn't expecting to see a big range of corals at Ningaloo: there would not be many that could have travelled south from the tropics against a north-flowing current, and a cold one at that. However, on arrival I immediately saw that this was not the case.

'Hey Barry, the current here comes from Indonesia; that old map of yours is wrong.'

This was obvious from the corals, and indeed all modern charts show the Leeuwin Current, now extensively studied, streaming south from Indonesia, hugging the coast for most of the way.

From the beach, I could see the reef edge, only a few hundred metres away, and I headed for it. A shallow break in the reef let me through, at which point I was in bottomless water with a modest current. I was a little wary, this being great white shark country, but the water was crystal-clear and I kept the reef edge in sight as I drifted along. Then I saw something very big coming towards me. I had been with whale sharks on the Great Barrier

Reef, but this was my first unexpected head-on encounter with one, so close that I was brushed aside by its massive body, leaving me to watch its enormous, jumbo-jet tail fin towering above me.

With one thing and another, I was impressed with what I saw of these reefs and vowed to come back and do a thorough study of its corals. Ningaloo, as locals love to point out, is Australia's largest fringing reef, a fringing reef being one that occurs close to the shoreline, as opposed to barrier reefs, which occur well offshore (although there are also many other kinds of reefs). Ningaloo is 240 kilometres long and covers an area of 700 000 hectares. It is deservedly known for its whale sharks, which now attract streams of tourists, and is also a corridor for the marine life of all south-western Australia, carried by the Leeuwin Current as it streams down the coast. In 2011 the reefs were given World Heritage status, something I had worked for, mostly by compiling the specialist evidence UNESCO needed: comparisons between Ningaloo and other World Heritage sites. Some enthusiasts like to compare Ningaloo with the Great Barrier Reef, but realistically there is no contest; the complexity and diversity of The Reef dwarfs Ningaloo in all ways. Ningaloo does, however, stack up against most other fringing reefs in other parts of the world, mainly because the water is very clear, due to an almost complete absence of coastal streams, and there is little human population pressure.

Most staff at the museum were excited by the idea of a survey of Western Australian reefs, and so over the next four years I worked in collaboration with the museum, visiting most of the major reefs of the entire coast. By this time Len had moved to another job, and Ed Lovell, a friend of Carden Wallace's, took his place as my assistant. We had some very interesting trips with the museum, the best being to the offshore atolls: the Rowley Shoals, Scott Reef and Ashmore Reef. To get to these the museum hired crayfish boats, fast fibreglass vessels used on the Houtman Abrolhos

Islands to the south. They did the job well enough although they lacked normal navigation equipment, this not being needed to catch crayfish. As it was well before the age of GPS, we found the Rowley Shoals by dead-reckoning, which is to say we roared out from Broome at 20 knots and kept going for more than 300 kilometres using nothing but a compass.

As it happened, we found Clerke Reef, one of the Rowley Shoals, because it was low tide and the reef's large white sandbar could be seen 30 kilometres away from our boat's flying bridge. That sort of tidal range in the open ocean was something new to me. At low tide, the lagoon inside the reef was perched so high above the surrounding ocean it looked like a giant ornamental pond. There were three narrow passages between it and the open ocean where water gushed out in a torrent, perfect for high-speed thrill dives and a mini-scale reminder of the deltaic reefs we'd seen during the Stoddart expedition.

With beautifully clear water, the Rowley Shoals made for excellent diving. We did a complete inventory of the corals there and found some new species, but the place was memorable because we happened upon a boat of Indonesian sea gypsies. These are little-known people who have their own language and spend their whole lives at sea, coming down the offshore reefs of Western Australia to fish every year, something they have done for centuries. I went aboard their boat with some presents, including my spare diving mask, something they'd never seen before. Their little boat had a charcoal burner on the back deck for smoking fish and bêche-de-mer, and wooden barrels of water were stored inside a tiny cabin. Incredibly, their navigational equipment consisted of a piece of string with bits of wood tied at specific intervals and an old kitchen clock. These would have helped give a crude estimate of latitude, but basically the sea gypsies found reefs by feeling the presence of refracted ocean waves, observing the reflections of reefs in the clouds, and watching seabirds.

I was angry when, many years later, the Australian government branded these people illegal immigrants, ending a lifestyle both unique and harmless. I wrote a strong letter of protest, but it was ignored.

In 1983 I made the first of several visits to the Houtman Abrolhos Islands, the southernmost reef complex of the Indian Ocean. From a biological standpoint these are among the most interesting reefs I've ever seen, a place where tropical corals battle it out with cold-water seaweed. The mix of corals we found was unlike anywhere else, with many species that were common there being rarities elsewhere, and vice versa. I found some new species during these trips and a few more that had already been described but were new to me.

But it was the seaweed that sometimes captured my thoughts, for seaweeds are seldom found in large quantities on healthy reefs. I saw that battles between seaweed and the corals are what these reefs are all about – a constant territorial stand-off that the corals must win afresh ever year. There are places where it's possible to stand with one foot in a mass of seaweed – *Sargassum* – and the other in coral. Any change, biological or environmental, however small, could alter the demarcation between the two. This is especially interesting because the stand-off changes seasonally, as the *Sargassum* comes and goes. Perhaps once a year the corals, along with the herbivorous fish they provide homes for, gain a temporary advantage, but this is short-lived, for it changes back as the *Sargassum* regrows. Just occasionally, larger coral colonies can be seen peeking through *Sargassum* beds in places where the wave surge is strong enough to clear away some of the algae and give the corals gasps of sunlight. If this occurs over large areas it can look comic, because the *Sargassum* seems to remain stationary while the corals appear to move back and forth.

The human history of the Abrolhos is as gruesome as it gets. The island on which I first worked was the site of a ghastly succession of massacres of the crew and passengers of the *Batavia*, the flagship of the Dutch East India Company, by mutineers after it went aground there on its maiden voyage to Java in 1629. Even when I was there it was common for fossickers to find human bones, even bits of skulls, in the coral rubble.

On that trip we discovered that a tourist had recently died of an embolism while diving – right where we were working. When his body was recovered the only sign of physical injury was to his right ear, part of which was missing. Interesting that, for I'd just been playing chasings with an unusually large and aggressive cuttlefish. It would come towards me in a threatening posture, tentacles upraised, its sides rippling with iridescent colours, and I'd chase it backwards. Back and forth we went until it got bored and swam off in a puff of ink. This left me wondering: had it grabbed the tourist by the head and bitten his ear? Being gripped by the tentacles of a big cuttlefish might well have been enough to send a novice bolting for the surface in panic, and to death from an embolism.

In another unexpected encounter, I was photographing a coral deep on a vertical reef face when I felt a tap on my shoulder. I turned and came face to face with a young sea lion, looking inquiringly at me. I needed that photo, so with a wish of apology I gave her a burst of bubbles from my regulator and turned back to my coral. *Tap tap* – she tried again, with her nose. She was very pretty, with soft baby eyes and beautiful long curved whiskers. I gave up, it was playtime. I whirled around in the water, releasing a spiral stream of bubbles, which had her cavorting like a child, which she presumably was. She accompanied me up to the surface, then back to the beach I had come from, and we both sat in the shallows for a chat. Soon other sea lions started arriving, until about thirty formed a wide arc around me, grunting and jostling. Sometimes they put their heads underwater to look at me, and sometimes they pointed their noses as high as they would go for an aerial view. That endearing moment ended abruptly when a big bull came charging in. I beat a hasty retreat; the bull clearly had no intention of sharing his harem with me.

One trip I made to the Kimberley coast, far to the north, was an eye-opener, as the tidal range there is one of the biggest in the world, peaking at over 14 metres. The ocean is shallow and constantly churning, so the water is always muddy. The coral gardens, high and dry at every low tide, are a spectacular site. They're surprisingly diverse, with a strange mix of species, many having weird growth forms that don't occur elsewhere, making some species difficult to identify. As the gardens bake for hours under the searing tropical sun, their temperature must go far above any recorded elsewhere in Australia, and perhaps even higher than the water temperature of the Persian Gulf, which holds most such records. My mind was overloaded with questions when it came time to leave: I longed to go back, and still do.

Corals exposed at low tide on the Kimberley coast.

The study of Western Australian corals went well, although (and this was something I could hardly admit to myself at the time) I felt a growing unease with the taxonomy I'd spent so many years building on the Great Barrier Reef. Most of the species I found in the west were more or less the same as those of the east, but many had some sort of taxonomic problem. I found myself living with an element of guesswork, quite unlike how it was on The Reef. There, once I'd worked out what a species was in all its many growth forms, I was confident of recognising it anywhere. In the west that confidence kept eroding, for the basis on which I separated one species from another on the The Reef now seemed doubtful.

I started to question my earlier work, not a happy thing for a scientist to do. These problems were soon to get worse, when I began working in Japan.

I didn't quite abandon the Great Barrier Reef at this time. For a start I had to reassure myself that I hadn't made the taxonomic mistakes that my work in Western Australia suggested. (I hadn't; Nature was playing a trick, one I will later describe.) On top of that, new outbreaks of crown-of-thorns starfish were causing havoc. Occasionally there were newspaper articles about them, but there was none of the interest that there'd been in the '60s and early '70s. The public had become inured to it all.

Had there been an inquiry of the sort envisaged a decade earlier, I could have been part of it, for I obtained a doctor of science degree, for my work on corals, from the University of New England in 1982. My graduation would have been the fifth time I crossed the stage at a degree-conferring ceremony, but after much soul-searching I decided not to go, because the thought of receiving yet another degree felt wrong when the only accolades that mattered to me – Noni's – had fallen silent.

That sounds selfish, I know, and I later regretted it, especially when the chancellor sent me a long handwritten letter expressing his personal regret at my absence.

A brief return to the Solitary Islands, 1986.

Travels Abroad

Trouble in Japan

By 1984 an interesting picture had emerged from all this work on Australian corals. We saw that on both coasts, species richness decreases in an orderly sequence from the northern tropics to southern temperate latitudes. That was hardly an earth-shattering revelation, but when this sequence was superimposed on patterns of temperature, currents and the occurrences of reefs, the results said a good deal about reef ecology. The question that naturally arose was: do similar patterns and correlations apply to the Northern Hemisphere – from Indonesia to Japan? I had been pondering this very question when I received a phone call out of the blue.

'Dr Veron? I am Dr Shirai. Please come to Ishigaki Island and report on the corals there. All expenses paid.'

I knew roughly where Ishigaki Island was, just north of Taiwan, at the southern end of the Ryukyu Islands chain. The timing and location could not have been better.

'Okay,' I said. 'When do you want me to come?'

'As soon as possible. Please bring personal diving equipment, camera and suit. We have tanks and all you need. We would be greatly honoured by your visit.'

A few weeks later I was on a plane to Ishigaki Island. As it circled around to the airport I saw that the island had a mountainous spine down the middle, a narrow coastal plain on all sides covered with rice paddies, and an almost continuous line of fringing reefs. A perfect place for diving.

I met Dr Shirai as soon as I walked down the aircraft steps. He was tiny, with an intelligent face and a happy smile, and was immaculately dressed, as if he were going to a business conference.

'So sorry the mayor is not here to meet you,' he said.

His English was excellent. I must have looked a bit surprised, because he explained that a mayor in Japan was the godfather of all things, not just a political leader. Then he looked a little worried. 'How old are you?'

'Thirty-nine,' I said, wondering why he'd asked.

'Please do not be thirty-nine.' He frowned. 'Forty-five is better. Please come with me, we have a meeting in the airport reception room where you will meet the press. Very important.'

A press conference about me? What's this guy on about?

The reception room turned out to be full of bright lights and big television cameras; there were many technicians in overalls, reporters with notebooks, and important-looking men sweating in their perfectly fitting suits, just like Shirai's. Nothing like this had ever happened to me before. Coral taxonomists just don't get this sort of treatment.

Someone welcomed me in faltering English, then turned to a camera and started a long speech in Japanese, which of course I didn't understand. This was followed by another welcome by someone else in a few practised English words, and then another long speech in Japanese. On this process went, again and again. Cameras, not to mention floodlights, were on me almost continually as one immaculately dressed man after another bowed formally and eagerly shook my hand. It seemed that I was some sort of celebrity. What was going on?

After two hours, during which I understood nothing, I must have looked as bored as I felt, for Shirai chipped in. 'Please excuse all this. It is the Japanese way. We are greatly honoured by your visit. We invite you to a banquet in your honour. Very important. But first I will take you to a beautiful guesthouse reserved for your visit. Did you bring a suit? No? Then I will have one delivered.'

Is he kidding? The last time I wore a suit was at my wedding.

The guesthouse was indeed beautiful, as Shirai said. It was built in traditional mainland-Japanese style and stood alone on a headland surrounded by trees. I had the whole place to myself, to be looked after by 'Papa-san and Mama-san', an elderly couple who spoke no English. But no matter; they certainly seemed a friendly pair.

The banquet, held in a large, tatami-mat room on the top storey of a modern hotel, was another novel experience for me. A circle of low, polished, heavy wooden tables were covered with platters of traditional Japanese food. There were no women, just middle-aged men, about thirty in number, all dressed as I now was in a grey suit complete with tie, tie pin, and a fake breast-pocket handkerchief. Sartorial splendour isn't high on my list of achievements; I'm sure I looked every bit as uncomfortable in my suit as I felt.

I was placed with Shirai on one side and the mayor, who had just returned from Tokyo, on the other. The mayor must have been about seventy years old and had a heavy build and a large, kindly face. He had bowed surprisingly low to me, and shaken my hand warmly. I took an instant liking to him, or at least to his smile. He spoke no English but we both knew a little German, his presumably from the war, so we got by with basics.

Japanese food has long since gone international, to great acclaim, but back then I'd never even seen a Japanese restaurant. I'd heard, to my horror, that Japanese ate fish raw, and there it was – in all manner of different colours, most wrapped in seaweed, with rice. I decided to start on something less challenging.

A small pile of green stuff on the side looked relatively harmless, and so my first mouthful of Japanese food was straight wasabi. It seemed an age before I could breathe again, much to the concern of the mayor.

Shirai was the only English speaker in the room, as far as I could hear. He chatted with me on occasion, explaining the food and showing me the way sake was drunk. As soon as anybody in the vicinity emptied their cup, all were refilled together. That way everybody got drunk at the same rate, everybody except me, that is, who apparently had a greater tolerance for alcohol than the Japanese. This seemed to give me a certain status, as did the ease with which I handled chopsticks (we used them at home) and sat on the floor – *gaijin* (foreigners) were not supposed to be able to do such things.

After the banquet had concluded I was chauffeured back to my guesthouse to the waiting arms of Mama-san and Papa-san, rather under the weather and wondering what the whole show had been about.

The following day there was no diving; the boat wasn't ready, I was told. Instead I was asked to attend another banquet in my honour, this time as the guest of the Director of Fisheries. This was held in another hotel, but seemed to me to be an exact rerun of the previous night – the same circle of men, no women, and the same food. The day after that the boat still wasn't ready, so Shirai collected me in his car, an enormous black Lincoln Continental, for a sightseeing tour. He was so small he had to sit on a large cushion to see over the dashboard. I'm only a bit over average height in Australia, but according to the locals I was the tallest of the forty thousand people on the island.

That night there was a 'formal dinner', which was more of the same except that this time Dieter Kühlmann, the curator I'd met in the Berlin museum nine years earlier, was there. Dieter had just arrived from Berlin to join me for a couple of weeks' study of the foreshores. He also did some diving with me eventually, as did Shirai and his assistant, but mostly I saw little of Dieter. Nevertheless, he was welcome company, particularly at the banquets, which never seemed to end.

For day after day our diving was cancelled on one unlikely pretext or another, but the banquets continued, always with the same men in the same suits. This must have been costing a fortune. Furthermore, Mama-san and Papa-san's television was almost always on and I'd see myself in every news broadcast. My photo was on the front page of nearly every newspaper. Why? It wasn't as if I did anything, and who would care anyhow? Shirai said it was just the Japanese way and to forget about it. This sounded increasingly phoney.

One day, when out walking, I was warmly greeted by a local dive shop owner, and so when yet another day's diving was cancelled I stole away from the guesthouse with Papa-san's ancient wooden wheelbarrow, went to the dive shop and borrowed a couple of tanks. The nearby reef flat was about a kilometre wide and the tide was in, so I swam to the outer edge, photographed corals all morning and then swam back, a journey taking several hours. Shirai thought this terribly dangerous. It wasn't, but it did prompt him to make sure our boat, a fishing trawler, would actually be ready the following day.

Thereafter, if the boat wasn't ready ('Very sorry'), I would go off by myself. Every time the boat *was* ready, there were reporters to interview me, in Japanese of course, and one or two television cameramen were always waiting. I would again be on television and my photo would again be in the local newspapers. Not Dieter's photo, nor Shirai's either, just mine, and of course I had no idea what all the newspaper articles were about.

After a couple of weeks of this, the mayor arrived at my guesthouse with a huge bowl of fruit and managed to say he knew that I was unhappy with my visit, so to cheer me up, would I please come to a special private dinner in my honour? I said I would; I always did. This time there were fewer men in suits, but there was a gorgeous girl, called Butterfly, seated next to me. Not only was Butterfly astonishingly beautiful, but she spoke perfect English, albeit it with a slight American accent. She explained that she was visiting Ishigaki Island from Tokyo. She had no idea why I was on the island except that it had something to do with diving. It was the most interesting dinner I'd had in weeks, even if it did take me a while to work out what her job actually was.

Two days were unusually interesting. On the first we went to the southernmost point of the island, where I embarked on a long swim to a bay that had no road access. The corals there were more or less the same as in other places until, abruptly, they were all white. Initially I thought this must have been the result of a crown-of-thorns starfish outbreak, but I soon ruled that out. Every coral was completely dead; starfish don't do that. I decided it had to be due to something dumped in the water, but then I remembered recent reports of mass bleaching on the Great Barrier Reef. This was the first instance of mass bleaching in Japan, and the first I had personally witnessed.

The second interesting day was a visit to a wide expanse of reef near the village of Shiraho, at the end of the island, close to the airport. The attraction was a lagoon in the outer part of the reef flat, clearly visible in the aerial photos we had. Shirai initially refused to go, saying that the villagers were dangerous; there'd been some sort of riot there. He gave in after I told him I would ask the dive shop owner to take me there if he didn't. So off we went, in his big black Lincoln Continental, which was a bit of a struggle because it didn't fit down most streets. However, sure enough, there was evidence of a riot everywhere, including a street barricade made of bits of houses, furniture and overturned cars. The village was deserted.

We hadn't taken scuba tanks that day because Shirai claimed he didn't want them in his beautiful car, so we set off wading. Shirai went knee-deep while his assistant took photos of him every few metres, with the village in the background. I kept wondering why as I continued on to deeper water alone.

The lagoon turned out to contain the biggest stands of blue coral, a living fossil, I'd ever seen. I had seen these corals many times, but never on this scale. I took lots of photos of them, later to be reproduced in my books and in magazine articles all over Japan.

After that, one day merged into the next, each with the same routine. Either the boat would be ready, allowing me to get on with diving, or I took myself off with Papa-san's wheelbarrow. By this time Dieter had returned to Germany. I carried on because I needed to complete the study for my own purposes, but I'd become so sick of reporters and cameramen that I bluntly demanded Shirai send them away, and in future not tell them when we were going diving. It was all very strange, and could not possibly be 'the Japanese way'.

The following morning things came to a head. The boat was ready, and so were the cameramen. After Shirai had made sure the cameras were on him, he came over to talk to me about the weather. I'd had enough. To show him the Australian way, I smiled sweetly at the cameras, picked Shirai up and tossed him over the side of the boat. *Splash.* I never saw him again.

The project was cancelled. I kept on diving and collecting and photographing corals, usually leaving my cameras on the beach under a towel when collecting. I soon became a tourist attraction, with one glass-bottomed boat after another hovering above me, their outboards making a horrible racket. I would give them the finger, until a honeymooner told me this meant nothing to the Japanese and showed me the hand motion that did. That worked.

One day I returned to the beach with my laundry basket full of corals to find that my cameras had vanished. I went to the guesthouse and tried to explain to Papa-san that I needed to call the police. He didn't understand, so I tried again. Then his face lit up; he led me into the kitchen and there they were, on the table, with my towel folded neatly beside them. The next day, I took the cameras on my first dive then went collecting, leaving them on the beach. The same thing happened; a group of children were 'helping' me. But why? I could only conclude that they were treating me as some sort of hero, as some of the local men did. Certainly they seemed to know something that I didn't about why I was on the island.

At that point the Director of Fisheries turned up and told me that he'd taken over the survey himself, and that I could do all the diving I wanted, except near the airport, where it was 'too dangerous'. When my work was all done – and he was very attentive, considering my barbarian ways – he flew with me to Tokyo, where we did some sightseeing and had some fabulous meals. Oh, how I had come to love Japanese food.

Of course I was very suspicious about all that had happened, so before leaving Japan I put together collections of newspaper articles about me and posted one set to the Australian embassy in Tokyo and the other to Katy Muzik, a researcher at the Smithsonian and presenter of natural history programs that were popular in Japan. I had met Katy in America and knew she'd once lived on Ishigaki Island, and that she spoke fluent Japanese.

The embassy made no response, but Katy knew exactly what was going on: a plan to increase the length of the Ishigaki airport runway to accommodate wide-bodied jets. That would involve extending it over the reef flat and Shiraho Lagoon. A large hill adjacent to the airport would supply the necessary rock. The mayor, the old man I'd trusted, had foreseen this and purchased the hill for next to nothing. When the airport development was proposed, the hill was suddenly worth $US70 million. The riot at Shiraho had been in protest against the development, peace being declared only after it was agreed that two foreign experts would make a study of the corals of the island and evaluate the worth of the reef that was to be destroyed. Shirai had allegedly been offered $US3 million to produce a report that vindicated the government's, and the mayor's, proposal. I was their reef expert, Dieter's role was to make notes and drawings of the foreshores. All the television appearances and newspaper articles had been fakes, always reporting me toeing the government's line. This hadn't fooled the locals, but that didn't matter to the mayor – decisions about such things were made in Tokyo.

Katy said I must write to *Asahi*, Japan's most widely circulated newspaper, saying why I had gone to Japan and what I had done. My brief article caused an explosion. All Okinawan newspapers carried a photo of me with headlines like SHIRAI ACCUSES VERON OF LYING. There was another riot at Shiraho. The following day the governor of Okinawa called an emergency meeting in Naha, the capital. I was accused of all manner of devious crimes, which, through a telephone linkup to my desk at AIMS that my accusers didn't know about, I could immediately rebuff. Television companies, especially NHK, had a field day. A dozen Japanese reporters must have come to Australia to interview me, and one morning NHK had me talking live on a breakfast program from Townsville, the first ever live coverage of a news event between Japan and Australia.

Over the several years that followed I made many more trips to Japan, usually accompanied by Moritaka Nishihira, a professsor of marine biology at a university in Naha and a delightful roly-poly character whom I dubbed my little Buddha. He and I visited all the tropical island groups of the Ryukyus that had good coral, and then Kyushu. But my last dives in Japan were without Moritaka, for they were along the coast of Honshu, where corals were scarce and not so interesting for him. Doing this, I found the most northerly coral communities in the world, at the Tateyama Peninsula, south-east of Tokyo. I caught a ferry from Tokyo and stayed in a little inn on the western side of the peninsula, near where the corals were. There were not only living corals there, but also fossil corals, in an area where the land had been uplifted several thousand years ago. Naturally I went fossil-hunting and soon found a 3-metre-deep drainage ditch where specimens of species no longer found along the mainland coast had been dug up. After many hours of scrabbling around in the ditch I looked up to see a row of peasants watching me, clearly at a loss as to what I was doing. This was a great time to test my fledgling Japanese; perhaps predictably, they understood not a word.

The morning of my last day at Tateyama I decided to go for a swim. There were no sharks of course – Japanese love shark-fin soup – the only thing to be wary of were ships going in and out of Tokyo. I waved at a couple of ferries that passed me but only received blank stares for my trouble. Swimming back, it seemed that the trees around the lighthouse on the tip of the peninsula kept changing, and then I realised I wasn't getting any closer. I was being swept around the headland by the current – next stop Hawaii.

That was a long day, and for a time I felt it might be my last. There was no way I could outswim the Kuroshio, but I thought – hoped – it would have a small coastal gyre as it headed into the North Pacific. So I swam across the current to where I thought this might be. It was rather scary but the sea was calm and the lighthouse was slowly coming into the right position. I had started out on the west coast of the peninsula and ended a long way up the east. It was getting dark by the time I walked back to the inn, hungry and very tired. That little swim was not a good idea but the study was definitely worthwhile.[14] It's one that can never be repeated, because the living corals have all been dredged and the fossils are now under apartment buildings.

I look back on my time in Japan with a good deal of wonder. Over the months that followed my stay on Ishigaki Island, both the mayor and the governor lost their jobs, but the battle over the airport raged on for ten years. In the end, Shiraho Lagoon was proclaimed a national monument, although I'm not sure what conservation value that had.

I kept returning to that strange country for years, to study, for conferences, and sometimes I was invited for ceremonial occasions. I don't know what happened to 'Dr' Shirai (the Dr bit was fake, or so I was told) but I did revisit Shiraho, where I was warmly welcomed. When I was leaving, a local took me to the town garbage dump to show me the remains of a burnt-out Lincoln Continental.

In 1992 I published a monograph on the corals of Japan.[15] And in 1995 Moritaka and I – mostly Moritaka's doing – produced the lavishly illustrated, rather expensive book *Hermatypic Corals of Japan*, in Japanese.[16] All well and good except that I had become familiar with Japanese corals in finest detail, and troublesome aspects about how closely related species could be distinguished were always cropping up. It was the same feeling of disquiet I'd felt when working in Western Australia, although in Japan the combinations of species that caused me trouble were mostly different. Once again I had to do some diving on the Great Barrier Reef to reassure myself that I hadn't made mistakes.

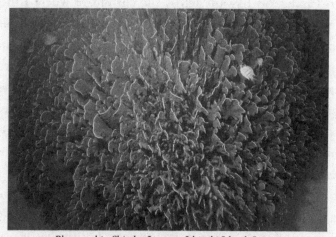

Blue coral in Shiraho Lagoon, Ishigaki Island, Japan.

Pacific forays

The earliest scientific account of the distribution of corals on the Great Barrier Reef was by John Wells, who originally described the dropout sequence of coral genera from north to south.[17] Nothing surprising in that, but what was interesting was that tropical corals did not have temperate replacements, as have almost all other marine animal groups. Why not?

In 1984, with the last volume of *Scleractinia of Eastern Australia* published, I made more trips to the ends of The Reef in between visits to the west coast. When satisfied that I'd filled in the main gaps in my records, I plotted the results and was almost dismayed to find that the dropout sequence for species was even more orderly than it was for genera. The far north of The Reef, where species numbers were low, was an exception, due to the turbidity and strong currents of the Torres Strait, an unhealthy combination for just about everything.

The sequence begged further questions. Did all groups of corals drop out at the same rate? Was it due to the temperature gradient down the reef, or battles with seaweed, or the presence or not of reefs? If the last, what was it about reefs that permitted a different diversity? The first question at least could be answered without further ado: mushroom corals (*Fungia*) dropped out quicker than the others, and so, to a lesser extent, did *Porites*. The interesting thing about *Porites* was that the colonies mostly became smaller in the south. This was unexpected because *Porites*, in experiments and in the geological record, are usually one of the best survivors when conditions get tough.

As work in Western Australia continued it became clear that a similar dropout sequence applied there, although much of that could be explained by the big gaps between reef areas. Both coasts had south-flowing currents and both had a temperature gradient from the tropical north to the temperate south. Both also had reefs, with a much higher diversity than nearby places where the temperature was about the same but there were no reefs.

Japan was the best place on the planet to take a better look at this question. The Kuroshio, the world's strongest continental boundary current, originates around the northern Philippines then streams north, past Taiwan, up the length of the Ryukyu Islands, finally turning east along the southern coast of Honshu. These conditions mean that there's no other place in the world where the relationship between temperature and diversity can be studied in anything like the detail it can in Japan.

These were the thoughts that bumbled around in my head when I first headed for Ishigaki Island, and they stayed bumbling around until I became determined to make a detailed study of them. Early in my travels to Japan I discovered that, being a nation of seafood addicts, the Japanese kept very careful records of ocean temperature, that being a good indicator of what seafood might be where. I could get some temperature records from central government agencies in Okinawa and Tokyo, but these were usually open-ocean records collected by ships of opportunity – mostly passenger liners and freighters. The records that mattered as far as I was concerned were those taken inside reef areas, right where the corals grew. These were kept by small research stations or local government agencies, which seldom shared them with anybody. To conquer this I developed a symbiosis with a couple of Japanese volunteers who accompanied me. The records, which the Japanese never published, would be politely handed over to me, 'the foreign expert', while my buddies did the talking.

Over the years we worked on about ten islands in this way, some in remote places. Each time, I would make a copy of whatever temperature records there were, then go diving.

This study confirmed the widely held belief that the minimum temperature for reef development was 18°C. That sounds simple enough, but why should it be so? What I found was that about a quarter of all corals could tolerate a temperature of 10.5°C, and a half tolerated 14°C. Even so, the old notion that reefs did not occur where the temperature went below 18°C held true, almost.[18] They won't grow if the temperature falls below 18°C for more than a few weeks. Corals can tolerate a much lower temperature, but a key factor is that they need at least 18°C to grow fast enough to provide homes for the herbivores that keep seaweed in check. It's a matter of ecology, not temperature tolerance.

It is now popular opinion that climate change will allow corals to disperse to higher latitudes. That's a sexy notion but it's been overplayed. Most reported latitudinal range extensions are actually due to improved identification skills or to the chance discovery of a species that wasn't previously recorded. That doesn't necessarily mean it wasn't there before. Some may come and some may go over geological time – the normal way of things for most organisms living at the extremes. Certainly, distributions are likely to change a little, but it will be change to distributions that have always been changing, and that's a hard thing to attribute to temperature or anything else, including differences in light regimes and mechanisms of reef erosion.

We may see corals appear on the north coast of New Zealand, as it may now be getting warm enough and they were there in the geological past, even building small reefs. It is more a question of getting the right current to take coral larvae there.

Today there are hundreds of magazines and dozens of books devoted to the underwater world of coral reefs, but in the early 1980s there was almost nothing. I'd been publishing my findings in scientific journals and monographs, but such publications conveyed little to the public, and nothing of the biology and beauty of reefs to anybody. So I decided to write a book for lay readers, which became *Corals of Australia and the Indo-Pacific*.[19] This was no small undertaking as it needed thousands of photos, making it expensive to produce. Angus & Robertson, at that time Australia's foremost publisher of natural history books, welcomed my proposal and so I began the long task of putting together a volume that ended up rather larger than intended. It had a lot of information that would have been new to most readers, including scientists.

I also had an ulterior motive: Noni had always wanted me to write a 'beautiful' book about corals and I wanted to dedicate this one to her. As a frontispiece I included a distant underwater photo of her taken by Ed Lovell on her first dive, at Keeper Reef.

Corals of Australia and the Indo-Pacific was published in 1986, and in 1987 I was privileged to be awarded the Whitley Medal for Australia's best natural history book – Noni would have loved that. As far as I know, the book contained the first photo of a bleached coral, something of little interest then, but prophetic of times to come.

The *Scleractinia of Eastern Australia* series and *Corals of Australia and the Indo-Pacific* gave both AIMS and myself a measure of international recognition. At least AIMS thought so – I was promoted to the top rung in 1987, the only promotion I ever got that didn't involve winning some sort of argument with a bureaucrat.

With my work on Japanese corals well under way, I was delighted when an opportunity came to make a study of the corals of the Philippines, where the Kuroshio starts. This was paid for in a rather unusual way: Ed Gomez, then professor of zoology at the University of the Philippines, had attended a coral taxonomy workshop that Carden Wallace and I had given in Phuket, Thailand, and he was keen to see this work extended to his own country. Being a man of great initiative, Ed had obtained a large grant from the US State Department for me to give a lecture at his university. When I arrived in Manila I went straight to Ed's office, where I saw a pile of pesos covering half a table.

'Good to see you, Charlie,' said Ed with a grin on his face. 'This was to be for your lecture,' he went on, gesturing towards the pile of banknotes. He then carefully pushed the pile to the other side of the table, except for a single note which he ceremoniously presented to me. 'This is for your lecture', he said, 'the rest is for your study.'

I never did give the lecture. I used the money to travel all over the Philippines with Gregor Hodgson, an American coral biologist who lived in Manila and spoke fluent Filipino, as well as a couple of regional dialects. I felt a twinge of guilt about not giving the lecture that America had so generously paid for, but Gregor and I did produce something much more valuable: a publication about the corals of the Philippines.[20] It was nothing like the detailed monograph it could have been, but we wanted to be careful not to tread on the toes of Professor Francisco Nemenzo, who was nearing the end of his days after spending a lifetime studying Philippine corals and producing a string of publications about them.

Our work involved some very interesting diving, all in small boats using scuba tanks that Gregor was always able to scrounge, even in small villages. One dive was more memorable than most: we had headed out in a canoe with an outrigger and tiny outboard motor. The driver spoke no Filipino, although he seemed to understand that he was to anchor – he had a length of rope tied to a chunk of rock for this purpose – and wait for us. When we finished our dive he was nowhere to be seen. This was no great problem because there were boats everywhere and we literally hitchhiked back to the village we'd set out from. When he saw us there, our boat driver nearly collapsed with fright – he thought we were ghosts. He'd never heard of scuba diving, and when we hadn't surfaced the poor guy assumed that we'd drowned and was terrified he would be accused of killing us.

Although I thoroughly enjoyed the Philippines, particularly the small towns and villages, I was continually shocked at the poverty, especially the number of destitute children. Gregor had to stop me giving away all our money; he also had to stop me getting into disputes with people who were treating animals with appalling cruelty. I was much bigger than most Philippine men, but Gregor insisted they had knives and would not hesitate to use them. I was also distressed by the plight of young girls. 'They're either married or for sale,' Gregor said, for a girl could earn more in one night than her father might in a month. They were usually taken back into their otherwise devoutly Catholic families once they'd reached their use-by date, early twenties at the latest, but after living in Manila hotels for years, and probably having contracted the diseases of their profession, a poor village life, perhaps without the prospect of a husband, must have seemed like a prison.

I saw the same thing throughout Indonesia, Thailand and Vietnam. The culture that these people had built up over a thousand years was being trashed wholesale by Western and Japanese wealth. Worse were the movies the villagers knew about, featuring live rape, even murder. I lived with locals in all these countries and had a glimpse of real life, so different from that served up by tourist operators.

For many years afterwards, I kept in contact with the students and university staff I met in Asia, often helping them to get jobs or scholarships. No trouble for me, life-giving for them. I was also able to repay Ed Gomez for his support of our study. Ed had founded the Marine Science Institute in the north-west Philippines. It was an impressive building, the likes of which did not exist in any other developing country. Then a Taiwanese company started to build an enormous cement factory right next door; it would have destroyed the institute. When I next saw Ed he was in bad shape, having fought against the factory with all he could muster, to no avail, and he believed his life was seriously in danger. I wrote to President Marcos, on AIMS letterhead, giving myself the status of grand professor of all things marine. I said I felt obliged to inform him that he would be shocked to hear of the imminent destruction of his precious institute (which he presumably neither cared nor knew anything about). He must have been in a particularly dictatorial mood, for he ordered the immediate closure of the factory. The whole exercise would have taken me an hour at most, and AIMS even paid the postage.

After the Philippines, Gregor and I went on to Vietnam to work on corals with local experts. Driving up the coast road at that time was like driving through a *National Geographic* magazine, with one rustic picturesque scene opening onto another. We spent three weeks on the central coast, working out the corals and comparing them with those of the Philippines on the other side of the South China Sea. Our journey ended in Hanoi, at a time when the whole country was jubilant about America's decision to normalise diplomatic relations. I found it hard to see what these people were so happy about given what America, and Australia for that matter, had done to them.

It was also a time when the US was lifting its embargo on White House tapes made during the Vietnam War, and I spent a surreal half-hour listening on the radio to the deep gravelly voice of Henry Kissinger telling Richard Nixon that 'nukin' the fuckers' wouldn't go down well with the American people. From my hotel window I could see Ho Chi Minh's mausoleum, right where the bomb would have been targeted.

I left my hotel to get a bite to eat. People were dancing in the streets in celebration, reminding me that many years ago I only just dodged being conscripted to come to their country to help kill them.

Shortly after the studies of the corals of the Philippines and Vietnam were finished I had an interesting sojourn in Vanuatu, which Terry Done organised. We had a large yacht, the *Coongoola*, perhaps not the fastest of vessels, but its owner knew the country and kept his boat in good shape for diving.

This turned out to be a thought-provoking voyage for me. Swimming over the coral, I saw none of the problems of taxonomic detail that had plagued me in Western Australia, Asia and Japan. I swam on, getting a little bored with seeing so much more of the same.

What's here that tells me I'm not on the Great Barrier Reef? The corals are identical.

A subliminal thought almost surfaced, then faded away when I saw *Zoopilus*, a large delicate coral that looks like a Vietnamese peasant's hat. This coral is not found on the Great Barrier Reef. Later I pondered the ship's chart. Vanuatu is just the other side of the Coral Sea from The Reef, not far away as currents go. I thought about the role currents might play in evolutionary change. But it wasn't until many years later that this thought became clear and I was able to incorporate it in my concept of reticulate evolution.

All such matters went on hold when we reached the island of Tanna and walked up Mount Yasur, an active volcano and mega-spectacular sight. The crater is gigantic. Deep in its middle is a second, small crater which, every half-hour or so, exploded with the noise of battleship guns, sending streams of lava half a kilometre into the air like a giant Mount Vesuvius firework and shaking the grey-ash ground we stood on. We watched transfixed, until some hot ash came down a little too close for comfort.

Not long after we were there, a tourist died from a direct hit by a chunk of scorching rock, and the place is now often closed to sightseers. A pity; unless the volcano is in one of its angry moods, the risk in seeing it is surely worth it.

The *Coongoola*, slow but roomy, Vanuatu, 1988.

The Indian Ocean

Six contour maps showing the global distribution of coral genera were published between 1954 and 1985, the first being John Wells's, the last being mine. All vaguely indicate some sort of centre of diversity in the western Indian Ocean. Why so? Where did the corals come from? The Tethys Sea, an ancient seaway which, as I will describe, periodically included the Mediterranean and covered much of Europe? Or did they come from the Indo-Pacific's centre of diversity? Or had they been there as long as Africa has?

When I started working in the Red Sea in 1985 these were unanswered questions – or rather, nobody had asked them. The Red Sea has a well-known geological history. It is mostly very deep but has a shallow opening to the Indian Ocean. That shallow opening means that all the life in the Red Sea has been there for less than fifteen thousand years, because the sea level during the last ice age was low enough to isolate the sea, turning it into a hot, lethally saline giant trough about as homely as the Dead Sea is today.

I first worked in the Ras Mohammad National Park, at the

southern tip of Egypt's Sinai Peninsula, in 1985, making a detailed inventory of the corals there and shipping a large collection back to Australia. Tourists had started arriving in substantial numbers by then, and Egypt was thinking that some sort of hands-on management might be a good idea. The first job was to find out what needed managing.

As soon as I started diving, memories of Ehrenberg's specimens in the Berlin museum came flooding back, but much more poignant were recollections of my detailed studies of Western Australian corals, large numbers of which I was seeing again. Or more or less seeing again, for most were a little different, not surprisingly considering the distance between the two countries.

I revisited the park with Mary (whom you will soon meet) in 1996 and was horrified to see that it was overrun with tourists, mostly Germans and Italians. Tourists have pretty much trashed the park now, there being limits to the numbers such a place can take, but at that time the reefs were still in good shape. The government was encouraging conservation because the park generated more foreign currency than the Pyramids and Sphinx combined. Egypt also made a fortune from ships that ran aground there, as we immediately saw when we arrived. The rangers took us to see where a big Cunard cruise ship had hit the reef. The ship itself, by then in a dock, was badly damaged and listing heavily, but Mary and I, reluctant witnesses, had a hard time finding much damage to the reef at all. Nevertheless, Cunard was fined £8 million, which was pocketed by the Egyptian government, not the national park.

Although I had all the necessary permits to collect corals – the first and last the Egyptians would ever issue, so they said – I had to keep my specimens covered up, even when I was with a park ranger, because the locals, as well as most tourists then, were very protective of them. We left the park with hundreds of photos of the corals and another detailed collection. The corals in the region are now disappearing rapidly due to oil spills, land development and mass bleaching. However, I did describe several new species and don't doubt that there are more to be discovered before they disappear.

On another trip a few years later, rangers took me sightseeing. I wanted to see some Bedouins, who had led the same nomadic existence since biblical days. I was shown into a traditional, dark, carpeted tent, where I could vaguely make out a man sitting in the corner. He started talking about corals. There was nothing unusual about that until he asked me a very specific question. That stopped me in my tracks, and in all honesty I could only say that he should ask Bernard Riegl, a well-known coral scientist who knew all about the matter.

'I *am* Bernard Riegl', he said, sounding like he had a grin on his face. Was there nowhere I could go where my knowledge of corals didn't get tested?

On another trip, I visited an ancient Catholic monastery, the site of the burning bush of biblical fame. The bush looked a little careworn, though not particularly old, but of much greater interest was the Catholic library, the second-biggest in the world outside the Vatican and the home of hundreds of ancient icons. The icons were still in excellent condition due to the low humidity of the Sinai, and the fact that Napoleon had ordered his troops to leave the place alone. I'd heard that one of the monks there was an Australian and I was going to seek him out when I felt a firm hand on my shoulder.

'G'day,' said a broad Australian voice. 'Heard you were coming. I'll show you around in a sec, but would you mind taking a look at this bloody laptop of mine, it's driving me nuts. It won't talk to my scanner.'

And now I'm fixing a computer for a monk in a Catholic monastery in the Sinai desert.

It turns out that the Red Sea has around 340 species of coral, only fifty or so fewer than anywhere in the western Indian Ocean. So all the old contour maps had got that wrong. A few species are endemic but they are unlikely to have arisen in the Red Sea itself. No corals could have survived the high salinities the Red Sea reached when it was cut off during the last glacial cycle, and the timeframe is not long enough for them to have evolved since then. So those species must have recolonised via the straits and then gone extinct elsewhere. Most western Indian Ocean species, including these, probably crisscrossed the Indian Ocean many times. Be that so or not, they aren't just exports from the Indonesia–Philippines archipelago, as are the corals of the eastern Indian Ocean: many appear to have originated in the west, and perhaps some go as far back as the Tethys.

Of all the places on the East African coast I visited over the years, Zanzibar was for me the most interesting. It had a colourful history as an ancient trading port and, more importantly, a small, privately owned island with some of the most unusual coral communities in the whole Indian Ocean. Chumbe Island Coral Park, then owned and run by a dedicated German conservationist, Sibylle Riedmiller, was special. Sibylle had been struggling with local politicians for years, trying to get legal recognition of her park. When I was there, the President of Zanzibar had publicly rebuked her, claiming that corals were rocks, so what was the point of protecting them? I joined the battle on her behalf – successfully, I think. A year later Sibylle offered to give me joint ownership of the park, but I declined. Such a venture might have been interesting but would have plunged me into yet more battles, and my life was complicated enough as it was.

The evening before I left, I chatted to the rangers who looked after the park. They were a small, dedicated group, and having a visitor who knew about corals was a big occasion for them, so we went on talking into the night. They told me of their work and revelled in the interest I took. I felt I should spend more time doing such things; it was a small thing for me, but so important for them.

Not long after I first worked in the Red Sea I made my first trip to Madagascar. It wasn't a particularly successful trip because I was on my own and in those days scuba tanks were hard to come by. Nevertheless, I dived in many places on the west coast, hoping that one day I would return.

It wasn't until 2005 that I had that opportunity, along with some colleagues. The trip was run by an American conservation organisation, which, with a curious lapse of judgement, chartered the *Inga Viola*, one of the most unusual boats I've ever been on. She was a 1932 Danish fishing boat but looked more like an old Chinese junk as she was made from curved planks that formed a rather beautiful sweeping deck, which unfortunately leaked profusely whenever it rained. A white box-like wheelhouse adorned the rear of the deck. The skipper, the only crew member, was an Englishman even older than his boat. Most interesting of all, the *Inga Viola* was powered by a 1928 diesel engine that by rights should have been in a museum. But there it was, with two cylinders, each about 2 metres high and looking more like a pair of microbreweries than an engine. The one used for starting needed kerosene poured into it from a bottle and compressed air from a scuba tank; it was such a complicated procedure that it took half an hour to get through – if nothing went wrong, which it almost always did. It was agreed that if needs be, meaning if the skipper died, I was to be the boat's 'engineer', because of my interest in engines. He gave me a five-minute lesson.

Long live the skipper. There's no way I'll be able to get this contraption going without him.

We loaded our stores and out we chugged at top speed, about 4 knots. Not surprisingly we broke down on our second day, or rather one of the cylinders did; a con rod broke. The skipper, not in the least fazed, unbolted the sump under the disabled cylinder, removed the con rod, wrapped a diver's weight belt around the crank shaft to balance the weight of the disconnected piston, and off we went again, *chug chug chug* at a new top speed of 2 knots. Not only were we dependent on that one cylinder for propulsion, but it also drove the scuba compressor and the generator; we couldn't even operate the ship's radio without it. God help us if we needed a mayday.

Mechanical problems aside, the diving was good and the offshore reefs were in excellent condition, but as far as I was concerned, the boat was a floating coffin; it had no chance of getting away from a cyclone, and it was cyclone season. With an uncharacteristic concern for safety, I got off the boat after a couple of weeks and worked on my coral collection in a guesthouse in Nosy Be, usually surrounded by inquisitive locals. While I was there a cyclone threatened, prompting me to start planning a rescue mission, but fortunately the cyclone missed the boat, which by chance was in a protected inlet still with the others on it.

'We didn't hear of any cyclone,' the skipper later said. That would have been because the cyclone had veered to the north and his radio wasn't on since the engine wasn't running. The following year another cyclone hit the same place and the mighty *Inga Viola* went down at anchor. I hope someone salvaged the engine.

Curiously, we recorded more species on our first dive than the French had during the years they occupied the marine station at Nosy Be. This made me wonder just what they'd done with their time: enjoyed their wine and cheese perhaps? By the time we'd finished our work we had confirmed that Madagascar had the highest diversity of corals in the western Indian Ocean, something we'd expected from previous reports but which had never been proven.

Back in Antananarivo, the capital, a rebellion broke out the day before we were due to depart and the airport was closed. We had no way of getting off the island, try as we did, even looking out for a passing yacht. However, the rebellion was peaceful, apparently enjoyed by all, and it allowed us to see something of the island. Most interesting of all were the fascinating animals the island is so famous for and which live nowhere else, including several species of lemur and chameleons. Madagascar is one of the poorest countries in the world, so slash-and-burn land clearing goes on unabated. All the forests, including some which were national parks, had been heavily logged; the only tree I saw that was thicker than a telegraph pole was on a privately owned patch of land, and had its own armed guard. That one tree would be worth several months' income for a would-be tree thief. The outlook for that wonderful place is grim indeed.

Unfortunately, although I have studied its corals in museums, I've never been to Chagos, in the middle of the Indian Ocean. This must be a special place for marine life, because it's the stepping stone that links the western Indian Ocean with the east. It is an extensive atoll complex, so perhaps it's not surprising that the three studies made of the corals there yielded very different accounts.

In preparation for a book I was planning, I made a string of field trips across the Indian Ocean in the late 1990s. These included a study of the corals of the northern Seychelles, and an all too brief look at those of the Maldives, but I did a thorough job in southern Sri Lanka and western Thailand.

Having done so much work on the Western Australian coast and seen the role that currents and temperature play in determining coral diversity there, I became interested in the opposite – in isolation. There are many places to study isolation in the Pacific, but there are not many in the Indian Ocean that are distant from major coastlines and the influence of strong coastal currents.

The Cocos (Keeling) Islands form a true atoll about halfway between Sri Lanka and Australia. I joined a Western Australian Museum trip there in 1994 out of curiosity about the corals that might occur in such isolation, and also because of the place's unique history. There are many islands making up the atoll, one inhabited by about five hundred ethnic Malays, mostly Sunni Muslims, and another by a small number of Europeans. At least that was the case when we were there; now there are five hotels. The Clunies-Ross family had occupied the atoll since the early nineteenth century and ended up running their own feudal government there, even minting their own currency. Not surprisingly, Australia took offence at this and put an end to it in 1978, so by the time we turned up, no family members were in residence. The original 1888 mansion, Oceania House, an extraordinary sight for such an out-of-the-way place, was empty but as the front door was wide open I had a peek around the ground floor. It was very impressive, with tiled floors and an ornate bar complete with glasses, full decanters, and bottles galore. The bar looked much like a traditional English pub, except for a dozen or so hens that wandered in and out, happily clucking away.

The locals there were obviously a law-abiding lot. The one and only policeman on the island told me there'd only ever been one misdemeanour – someone had helped themselves to a little money from the shop cash register, but that was many years ago.

One day a German photographer turned up and asked us to drop him off on one of the small islands. He took some underwater photos of the wreck of the SMS *Emden*, the German light cruiser that had been fatally damaged by HMAS *Sydney* during World War I. He made a lot of money from those photos: why weren't we smart enough to think of that?

The corals were rather disappointing. The atoll lagoon had turned anoxic (devoid of oxygen) due to a rare failure of the usual tidal flushing, and there were no corals left alive when we arrived. But we dived a lot on the outer faces, which were in good shape. Biologically there were few surprises. The isolation of the islands was always going to dictate that the diversity would be low, and the presence or absence of species appeared to be a matter of hit-and-miss, rather like terrestrial island biogeography, about which theories then abounded.

Over many subsequent years we eventually made sense of the origins of eastern Indian Ocean corals, most of which are exports from the equatorial central Indo-Pacific. The key to these migrations is, as I've said, the capacity of coral larvae to make long-distance journeys, something we still have a lot to learn about, but which is fundamental to all coral biogeography and evolutionary theories. That corals have a great ability to disperse is taken for granted today, but before this was realised most theories put forward on the subject read like fairy tales, with no possible basis in reality.

A sad end and a lifeline

Throughout much of the mid-1980s Kirsty had been increasingly feeling that while I was getting on with my life – travelling, having all sorts of adventures, working, travelling again – she was stuck at home bearing the brunt of family responsibilities and unable to exploit her own talents, except when in a theatre production. And when she was, she had a long drive, at night, to wherever she was rehearsing. I knew this was unfair; it was no small matter.

Coincidentally, in 1988 the husband of a schoolfriend of Kirsty's who ran adventure tours as a hobby asked me to be the guide on a trip to the northern Great Barrier Reef for writers and photographers. As this was a perfect opportunity for Kirsty to see The Reef for the first time and to do so in her sort of company, I agreed. The trip, to take place the following year, would require some time to organise and in the meantime we took a long-planned holiday in Israel, arranged by our Israeli friends.

In those days it was still possible to travel anywhere in Israel and we did, even wandering freely around Jerusalem, which was soon to become impossible. Katie, whose health had continued to improve, enjoyed floating on the Dead Sea (as opposed to in it, due its extreme salinity) and we were all entranced by Masada, the mountain fortress where the Jews made their last stand against the Romans. Kirsty and I enjoyed the company of the Israelis, including new friends we made, but Kirsty felt that I was still working. One day I went diving at Eilat, Israel's tiny strip of reef at the tip of the Red Sea, and received a hot reception when I inadvertently surfaced in Jordan. I also looked for fossils when we went into the desert. In that sense, I knew I was always working. I still am, and always will be; it's who I am.

Perhaps our return home highlighted the problems Kirsty was having with me: it seemed to her that I was never there for her, and it was true. From my point of view, no amount of work or travel seemed to stop the relentless decline I'd been in since Noni's death. Kirsty announced that she'd had enough; she decided to leave Rivendell, with me or without.

So in May that year we bought a house in town, and heartbreakingly I put Rivendell up for sale. Kirsty moved into the new house and offered me a room where I could come and go as I pleased. I tried doing that, and spent a couple of nights listening to the neighbour's dog barking instead of possums arguing, and smelling the neighbour's cooking instead of wattles. It felt horribly claustrophobic. I lasted only those two nights, then withdrew Rivendell from sale and went home.

Kirsty's move wasn't just about real estate. Our once idyllic marriage had seen such unrelenting bad times that it had become little more than a survival mechanism. I started to slide into bouts of my old pill-popping depression, taking the phone off the hook when at home in the evenings for fear someone would call. Buffer, my much loved labrador, was always at my side.

The trip to the northern Great Barrier Reef went ahead later that year. Kirsty's friend had chartered a 200-ton converted Scottish collier, the *Noel Buxton*, which seemed just about due for the scrapyard. Fortunately the weather was good, so she stayed afloat and we were able to see the far northern ribbon reefs without too many of our number getting seasick. We stopped off at some of my favourite places, including Raine Island, which we could only see from the sea, as by then it was under strict protection and my special-exemption status had expired. We kept going until we reached Mer Island, where I met again some of the locals I remembered from visits long before.

It was an enjoyable trip for all aboard, which included Issie Bennett. It was always good to catch up with her, and as usual we had many long talks. I was only just beginning to realise how the freedom I'd had to be in the natural world as a child had moulded my future, and I found myself talking to a kindred spirit on that score as well. It was only then, too, that I learnt it was Mrs Collins, my old teacher, who'd persuaded my mother to take me to see Issie and who'd later told Issie about my scholarship to the University of New England.

I think Kirsty, who'd come to know Issie, would have enjoyed the trip, but she had decided against going. Being married in theory but not in practice had left us with nowhere to go, and so we agreed to separate. Divorces are never nice things, especially after twenty-two years of marriage, and more especially after all we had been through, but if anyone must have one, let it be like ours.

The Family Law Court judge was an amateur actor and had been in a production with Kirsty.

'Charlie,' he said, 'you are aware that I know Kirsty personally; do you think this might influence these proceedings?'

'No,' I said.

'Kirsty,' he said, 'you are aware that I know Charlie personally; do you think this might influence these proceedings?'

'No,' she said.

'Well,' said the judge, 'you guys know what you're doing. Done.'

Kirsty and I then had a very sad dinner. It was all rather surreal.

There were no lawyers telling us what to do. Kirsty had her house and such investments as we owned; I had Rivendell and Buffer. It wasn't until much later that Mary, forever alert to social justice, pointed out that this wasn't a fair division. I had a career and a good salary; Kirsty had neither, for it was she, not me, who'd sacrificed a career for our children. I needed to remedy that, Mary said. I suppose these sorts of issues occur with most divorces, but Kirsty's and mine was unusual; we were able to settle such matters ourselves, and we remain close friends.

I first met Mary Stafford-Smith when she came to my office at AIMS to talk about her PhD project – the effects of sediment on corals, which she was doing at Lizard Island on a scholarship from York University. All I remember of that conversation is Mary telling me what she thought I should think of her project. At any rate, she did most of the talking and seemed to have things well under control.

A couple of years later I was sitting at a table in the AIMS canteen having lunch. There were several other people there, including Mary, who was telling engaging stories about life on Lizard Island, and in particular her encounters with Agro, the island's crankiest sand goanna. By the end of lunch she had me entranced. This was no small thing, as I'd been alone at Rivendell, shunning any contact with humanity, for a long time. I decided to invite her to dinner, then took a week to summon up the courage.

Shortly afterwards, Mary headed off to her mother's home in England to write up her thesis. I visited her a few months later and she took me to see Cheveley Park, the horse stud her family had owned and where she'd lived until she was seventeen. I was flabbergasted. The grounds, the trees, the main house and the hundred-odd horseboxes (stables), complete with red-tile roofs and clock towers, were staggeringly beautiful. There were fields of freesias and daffodils everywhere, and giant cedars of Lebanon, more than nine hundred years old.

Mary immigrated permanently to Australia in the latter part of 1990. I met her at Townsville airport, a small figure almost entirely hidden behind 120 kilograms of luggage on a trolley, which she, Mary-style, had talked herself out of paying any excess baggage for. My new life had begun, one that Mary was to transform into something I could never have imagined.

Early on, she decided that we should have our own seagoing boat. It needed to be a displacement hull, not a speedboat, and one that could be towed on a trailer by our Troop Carrier. After much looking we found the perfect design, a boat with a solid fibreglass hull and diesel engine. We also found someone who could build it for us in Brisbane. *Wanda* (named after the fish) was a terrific little boat, incredibly seaworthy if a little slow.

She proved her worth when Mary, having obtained a post-doc scholarship at James Cook University to continue her studies on coral, spent most of her time in 1993 at the Orpheus Island research station doing experiments. Her field work was done from *Wanda*, with me turning up at regular intervals, helping here and there and enjoying the working holiday.

This was a good period for my own work too. In 1992, with a monograph on Japanese corals published, I followed with another on coral biogeography, which brought together all that was known on the subject at the time.[21] Mary helped a lot with these, as well as several other studies I was doing. She had computer skills I could only dream of, and in return I helped her with my knowledge of corals and with technical aspects of her experiments. She was becoming a good critic of anything I wrote – we forged an extraordinary working partnership on top of a loving personal one. I had a guardian angel somewhere, even though it had a lousy track record.

Everything seemed rosy indeed until, in passing, Mary mentioned the subject of children. She wanted children! With all I'd been through, wasn't this the last thing I wanted? But after a rethink I realised I had to stop the tragedies of the past from dominating my future, so the ensuing conversation was brief. A short time later, Mary's pregnancy test was positive. Another new beginning had just begun, but in the meantime we had a journey to make.

Our Troop Carrier wasn't just good for towing *Wanda*, and Mary was keen to see the outback, especially after hearing of my travels there as a student. Back then I'd met up with several groups of Aboriginal people in very remote places and made a point of talking to the elders about their spiritual affinity with the land. I imagined I understood that, at least a little, for it seemed not so different from how I felt about my own world. And so in the winter of 1994, with Mary six months pregnant, we teamed up with her brother Mark and his partner Jenny, who reassuringly was a doctor specialising in obstetrics, and made a seven-week pilgrimage along the Canning Stock Route, which runs inland from the Great Victoria Desert in the south to the Great Sandy Desert in the north.

At the northern end of the Canning we farewelled Mark and Jenny and continued on by ourselves to the southern border of the King Leopold Ranges Conservation Park. We stayed there a week, camping under the stars among the many branches of the Drysdale River, at least 100 kilometres in all directions from any other human and perhaps where only Aboriginal people had ventured before. It would have taken months for me to get my fill of that lovely peaceful place. I sat and watched the river for hours: a freshwater crocodile might come and go, and if I was really still, all sorts of other wildlife came close by – jabirus and even a small mob of emus. I regularly cooled off in the river, freshwater crocodiles being harmless (something Mary took a degree of convincing about), then explored a little or did nothing. Unhappily, doing nothing wasn't what it once was. Thoughts persistently invaded my peace: I was losing the gift of not thinking, and felt much the poorer for it.

Our daughter Eviie was born in December, not long after our return. Following a brief period of anxiety on my part that it could all happen again, we had a time of celebration the likes of which I hadn't felt in many a year.

My sister Jan lived to see Eviie, but only for a couple of months. Fourteen years had passed since Noni's death, yet Jan was still dealing with her private sorrows by drowning them in alcohol. I went to see her in Sydney soon after getting back from our trip and it was obvious to me, if not to her, that she was in a bad way. I persuaded her to sell her house and move to Rivendell where we could look after her. She became terminally ill while with us and died after two weeks in hospital. She was fifty-seven. What a waste of a life. It was only in the last few years, after she'd moved out of the family home, that she had a chance to do anything she herself wanted to do.

Time and Place

Invading deep time

I first visited the Caribbean in 1992 with Mary, who had worked there several times before and had a good knowledge of its corals. On arriving in Belize, we went on a tour of Mayan ruins, then had a stunning journey along freshwater rivers that wound their way into the jungle. The trees were dripping with epiphytes I knew existed but had never seen. Then, joy of joys, we found juveniles of the cichlid fish that lived in our biggest aquarium at Rivendell. I could have spent months in that place, but, duty calling, we headed for Glover's Reef. There, Mary no doubt enjoyed giving the expert his lessons, but at the risk of sounding tedious there wasn't much to learn and I knew the species anyway. A patch of central Indo-Pacific reef the area of our kitchen is likely to have more species than the whole of the Caribbean combined. Thus it was not so much the coral species that interested me, but why they were so distinctive, and the story of how they arrived.[22]

The birthplace of modern corals doesn't exist today, for it was the Tethys Sea, a place with origins going so far back in geological antiquity – over 200 million years – that the face of the earth was then unrecognisable. Around 100 million years ago the continents of the Southern Hemisphere had started to relinquish their hold on Gondwana (of which Antarctica is the last remnant), so there were several east–west seaways circling the tropical world. The most important of all these as far as marine life is concerned was the enormous gap between North Africa and the place we now call Europe. At the end of the Cretaceous, around 70 million years ago, with dinosaurs contemplating their demise, the Mediterranean was connected with the Indian Ocean to the east and with the Atlantic Ocean to the west. Further north, with the northern movement of Africa unstoppable, the western Eurasian plate had buckled and sunk so that eastern Europe and the Middle East were also joined to the founding Mediterranean. This was the Tethys Sea, with ever-changing shorelines that usually extended from Africa, at that time still well south of its present position, all the way to Scandinavia.

With the southern continents colliding with those of the north, the earth continually shuddered. That meant strings of volcanoes erupted, which in turn meant carbon dioxide belched forth. The earth, by today's standards, was hot. With the southern Tethys, now called the Mediterranean, getting deeper and the northern Tethys, now Europe, getting shallower, a large expanse of high-latitude warm sea developed. This 'Super-Tethys' was reef-building country at its best, and the corals that built them abounded in both number and diversity. Like reef corals today these were powered by sunlight-capturing zooxanthellae, but these zooxanthellae, unlike those of today, were heat-resistant, having had millions of years of selection to become so.

Of course the Tethys, the birthplace of so much marine life, was doomed. By the Middle Miocene, around 15 million years ago, Europe was undergoing one of the greatest land reclamation projects in the Earth's history. As the map on page 214 shows, this reduced the Tethys to a seaway north of Saudi Arabia, but that too was being choked off as the Red Sea deepened to the south.

When that process was complete, all that was left of the Tethys was the Mediterranean, where there are no reefs and almost no zooxanthellate corals.

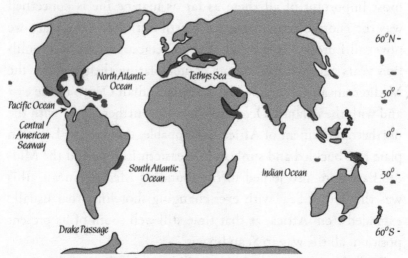

Continental positions during the Middle Miocene, about 15 million years ago. Surviving reef limestone is shown in dark grey.

Just how corals moved out of the Tethys attracted some bizarre theories before biogeographers realised that their larvae could make long-distance journeys. Some corals travelled east to discover east Africa and northern India, before the latter slammed into Asia. Some travelled west to reach the Atlantic, and some travelled nowhere and went extinct. This is a drastic simplification of 100 million years of the Earth's history, for that history was anything but simple. No doubt there were many migrations and many reversals back to the Tethys, but the main events relevant to this story were the extinctions, for carbon dioxide repeatedly acidified the oceans, and each time it did corals were decimated; some survived and some didn't.

Most of the corals that reached the Atlantic crossed it and so came to the Caribbean. That crossing was not as far as it is today, but more importantly it was warmer; the Atlantic wasn't the barrier it would be now be. Those corals that made the crossing found almost another Tethyan Garden of Eden, for the Caribbean was not isolated. There was no Isthmus of Panama, so the Caribbean was a passage that linked the Atlantic with the Indo-Pacific. And so the journey of the Tethyan corals continued, up and down the Pacific coast of Mesoamerica.

When I went for my first dive in Belize I couldn't help imagining I was diving in the Tethys (ignoring passing Plesiosaurs). All the species and about half the genera were new to me in their living state, and many were strikingly different from any I had seen before. This, I imagined, was what the cradle of modern corals might have looked like. To some extent this was true: the average geological age of Caribbean coral genera is twice that of the Indo-Pacific, but in the distant past the diversity of the former was far greater than it now is.

Several million years ago the gradual formation of the Isthmus of Panama started separating the tropical Pacific from the Atlantic, eventually leaving the corals on the Pacific side completely cut off from their Caribbean cousins. That would have left time enough for new species to evolve on the Pacific side, and indeed they probably did until disaster, in the form of the ice ages, intervened. The cold Humboldt Current, running north along the west coast of South America, wiped out most if not all corals of the entire eastern Pacific. On the Caribbean side conditions were generally better, for the corals were kept warm by the westward flow of water from the tropical Atlantic, just as they are today.

The history of all coral reefs is one of disasters and the Caribbean is no exception. Their greatest disaster was yet to come, caused by the ice ages, but not by temperature. To see this we must jump forward to about 25 000 years ago, when a massive ice sheet built up over much of North America. A large portion of the sheet (or shelf, as it is called) was over 2 kilometres thick. At that time several Milankovitch cycles kicked in and the sheet melted, and did so quickly. A glance at a map shows what this meant for the Caribbean, for the Mississippi River drains most of North America and even some of southern Canada. A volume of fresh water beyond imagination poured into the Gulf of Mexico and thence the Caribbean. The fate of the corals there is not known, but the event would have been catastrophic, perhaps wiping out most of them. The most likely explanation for their survival and subsequent return is that the corals hung on in the south-east, protected from freshwater by the incursion of tropical water from the Atlantic.

The corals of Belize that I saw would have had to struggle with the sea level changes, as all corals did, but I can't imagine they survived all that fresh water, which is lethal to corals. The corals there today must be immigrants, and geologically recent ones at that.

After leaving Belize we headed for Discovery Bay Marine Lab on the northern coast of Jamaica. Marine biologist Tom Goreau (1924–1970) had made the station famous, partly because of his historic studies of reef ecology (John Wells, who was anything but an ecologist, had a photo of him on his living-room mantelpiece), but also because he was the first person to use scuba (in the mid-1950s) for coral research. That's a long time before I did.

When we saw it, Discovery Bay had almost no coral; we dived mostly on bare rock. Mary had worked there before and was dismayed by the change. The destruction was well known and much studied, but the real shock came when we went through the black and white photos Tom had taken during his time there. The corals were once lush and abundant, as good as any we ever saw anywhere in the Caribbean. It was a stark reminder that such disasters can and do happen within a single human lifetime.

I have worked in many other parts of the Caribbean at one time or another. Some students claim they can recognise geographic variations in some species there. If so, good for them. I never could; at least, not remotely on the scale we see in the Indo-Pacific.

The fossil record of corals is by far the most telling of all animal groups because corals build reefs and the fossils these reefs contain give us windows back to the very beginnings of animal life. This makes reefs Nature's historians, places that can be dated and which keep track of changes in all the organisms they preserve, something they have been doing for hundreds of millions of years.

The history of all reefs goes in boom-and-bust cycles, but the end of the Palaeozoic Era, 250 million years ago, was a bust like no other. Every coral on our planet, along with so much else, went extinct because our oceans became lethally acidified by a massive spike in atmospheric carbon dioxide and as a result were almost devoid of anything that had a carbonate skeleton. These conditions took millions of years to subside, and it was many more millions of years before new groups of animals evolved to replace those that had gone extinct. This evolution included modern corals – the Scleractinia – which came from what were probably soft-bodied, anemone-like ancestors only very distantly related to the corals that built the reefs of the Palaeozoic.

From the mid-1980s I took a good deal of interest in what palaeontologists had to say about the fossil record of scleractinian corals, a time that spans the Mesozoic (the era of the dinosaurs) and the Cenozoic (our own era). With the help of my computer, such as it was then, I built a detailed compilation of that record, and this became the start of a very big problem, because a high proportion of Cenozoic families and some genera are alive today and these had to be amalgamated with the older, extinct taxa from the palaeontological literature. The result was a family tree, which I first published in *Corals in Space and Time* (1995), that looked nothing like that of its widely accepted predecessor produced by John Wells thirty years earlier. His had one trunk and main branches, whereas mine was coppiced, a tree of many trunks all growing from the bottom. Some of my branches were cut off, representing families that went extinct, while others reached the top of the drawing, representing families still alive. My tree made some European palaeontologists angry, this time because I was invading their territory. What right did I have to do that?

In 2007 I was delighted to accept an invitation to an international symposium in Austria on fossil corals. When I was asked to be the opening speaker I knew the organisers wanted to kick off on a controversial note, so I talked about my concept of reticulate evolution and then about my family tree of corals. This soon had some of the audience in the front rows shaking their weary grey heads and scowling at me: the organisers got their money's worth. However, most delegates seemed happy with what I had to say and several specialists later went over some of the finer points of the fossil record with me. The symposium was great fun and I appreciated talking with palaeontologists, whom I seldom got a chance to meet. I especially enjoyed the field trip, to a place full of Mesozoic fossils on the slopes of a massive reef they'd once built, the calcareous Austrian Alps. I doubt that the thousands of people who ski down these slopes every year realise they're skiing down the face of a coral reef.

One day, I hope someone will revise the old coral palaeontology catalogue published by John Wells in 1956, a daunting task that has been talked about for decades.[23] If they do, they'll hopefully heed what is being done with modern corals and use decision-making technology that can separate fact from opinion or, dare I say it, fiction. A program I have been using for the past fifteen years does just that. Every newly added taxon needs justification in the form of characters that separate it from every other taxon. This was the notion that most outraged the old palaeontologists: how dare I say that a computer was smarter than they?

I wrote a paper for them, laying out the facts of the matter. It was rejected, of course.

I give up. These old guys should too.

While I was attending a seminar at James Cook University in 1987, Russell Kelley, then a coral palaeontology student, passed me a coral – a small *Porites* – and asked me what it was. I knew that he knew what it was, so what was he on about?

He whispered that it was about 2 million years old, but it looked as if it had been collected yesterday. It was from the coast west of Port Moresby, Papua New Guinea. Enough said; we had to go there.

Today, Port Moresby is among the most dangerous cities in the world, and although back then it wasn't so bad, some of the surrounding districts were, and unfortunately they included the piece of coast we needed to get to. Nevertheless, Russell persuaded the University of Papua New Guinea to lend us a Land Rover and off we went, north-west from Port Moresby along a rough track through the forest. A couple of times some 'rascals', as they're called, tried to stop us, so we were relieved when we arrived at a small Roman Catholic mission, where one of the two inhabitants offered to take us to meet the local villagers. All was well after that, especially when a mob of children joined us, collecting coconuts for us to drink and helping to carry our fossils back to their village.

The fossils were, as expected, exceptionally well preserved and the place turned out to be the richest coral fossil deposit in the world. We had no means of determining the age of the fossils from radioisotopes at that time, but as the whole region had been stratigraphically mapped in detail we were reasonably confident that they were 2–3 million years old. That's not enormously old by most geological standards, but it does come into the category of time out of mind for most people. The fossils were of particular interest for me because by then I'd worked on the living corals of the region and so had a good basis for seeing what changes had taken place over that amount of time. With rare exceptions, fossil corals cannot be reliably identified to species, but this collection was exceptional because the corals had been encased in mud, which turned to stone, shielding them from deterioration. Extraordinarily, sixty-five species of fossil corals seemed identical to their modern counterparts, or nearly so; nine species were clearly different, and a further nine had almost certainly gone extinct.

When all this was tallied up in detail we concluded that the average geological age of the coral species we studied was about 20 million years.[24] That's a huge span of time compared with most animal life. Our study showed that scleractinian corals are a slowly evolving group, although nowhere near as slowly evolving as some 'living fossils', of which blue coral (that I'd seen in Shiraho Lagoon) is among the oldest, dating back more than 70 million years.

This study seemed satisfying at the time, and from a palaeontological perspective could hardly be bettered, but I had doubts that lingered. Our work was based on the morphology of a few time-eroded specimens of each species. I now believe that studies using DNA, which can give a significantly improved insight into the longevity of species, would indicate a much shorter time span.

Even though this trip resulted in my getting another dose of malaria, I would love to go back again and explore further. I have tried a couple of times, but it remains too dangerous.

Hell's atoll

In 1994 I had a chance to visit Clipperton Atoll, the only atoll of the entire eastern Pacific. By this time the notion that corals could disperse over very long distances was firmly established, something that had become obvious when I first plotted species distribution maps.

When writing *Corals in Space and Time* I was a little irritated to read a condemnation of what was then a novel proposal, by T.F. Dana, that the corals of the far eastern Pacific (the west coast of Mesoamerica) were recent immigrants from the central Pacific. I took sides, as much on behalf of Dana as the corals.[25] A few species, which appear to have gone extinct elsewhere, might have survived the ice ages in the east, but most are indeed immigrants that crossed the vast empty space of the eastern Pacific from islands in the west. That's saying something about the endurance of the larvae of these species, for these journeys must have taken months, the larvae floating on the surface of the east-flowing North Equatorial Counter Current. However, they took their zooxanthellae with them; they didn't go hungry, and several species probably cheated by growing on floating objects like pumice, spawning asexually, or releasing egg and sperm bundles during monthly lunar cycles rather than annual mass spawnings. The chances of colonising locations as remote as the Mesoamerican coast, let alone Clipperton Atoll, would have been extremely remote. Nevertheless, this is what happened.

Our trip to the atoll was organised by John Jackson, an American friend who was later to help with the publication of *Corals of the World*. We knew that Jacques-Yves Cousteau had visited the atoll in 1978 to make a film about its incredible human history, but he found so many ultra-aggressive sharks there that he, of all people, had to abandon diving. So on our first dive we entered the water *very* carefully. No sharks. In fact we didn't see a single shark the whole trip; Mexican shark-finners had done a depressingly thorough job.

As expected, there were very few species of coral on the atoll, but the few that were there made a thriving if singularly weird array of coral communities. I collected seven species of coral during my first dive, and that turned out to be the total inventory of all I saw the whole trip. I'd read several theories centred on the notion that reefs had to have a high species diversity in order to flourish or even exist. Not so: this atoll is built almost entirely of *Porites*, as were the very last reefs of the Tethys.

With not much coral work to be done I had time to have a good look over the atoll itself, which is shaped like an enormous doughnut. The interior lagoon is brackish and completely surrounded by the island proper, which has just one prominent feature, a 30-metre-high chunk of volcanic rock complete with a couple of caverns.

The place had an incredible history. In 1914 a Mexican garrison of about a hundred men, women and children were sent to the island to secure it as Mexican territory. The island had one other occupant at that time, a large, delusional African-American lighthouse keeper by the name of Victoriano Álvarez, who kept the rock for himself. After the outbreak of World War I the Mexicans forgot about the garrison and by 1917 all the Mexican men had died of starvation and disease. But amazingly, some fifteen women and children were still alive, all suffering from scurvy and the children severely deformed from rickets. Once they had lost the protection of their men, the women were enslaved by Álvarez, who took them, one by one, to his rock, raping and beating some and shooting others.

In a story with few rivals, the widow of the garrison commander and her maid managed to club Álvarez to death on the very day a rescue party, in the form of an American warship, arrived. The women left Álvarez's body to the orange crabs that abound on the island, and the captain of the warship kept his mouth shut lest the women be accused of murder. One of the children who survived subsequently returned to the island with Cousteau's expedition, as the star of his film *Clipperton: The Island Time Forgot*.

Much later I talked to Cousteau about this trip and much else besides, especially the discoveries he'd made and his endearingly heartfelt love of the sea. I never thought much of his movies, though they were groundbreaking for the time, but I sure took to the person, a deep-thinking and gracious old man. I would love to have known him better, it would have been such a privilege.

A different evolution

As I've previously noted, once I had a good grip on the taxonomy of the corals of the Great Barrier Reef, the more I studied those in regions elsewhere, the more I questioned the details of my original work. The problem wasn't just to do with geographic variation within species, it was that species appeared to converge in some places and diverge in others. I found this hard to explain to myself, let alone anybody else, and it alarmed me so much that I even considered abandoning my lifelong goal of creating a taxonomy that spanned the entire Indo-Pacific. Then suddenly the fog cleared.

Being a creature of habit, I always get up early in the morning and usually make myself a cup of tea before going to my study. On one such morning in 1993 it was business as usual until, half asleep and waiting for the jug to boil, a concept of evolution very different from Darwin's came into my head, unprompted yet fully formed. Well, almost fully formed – just how this evolution interfaced with Darwinian evolution took me several years to think through. Revelations such as this are popularly called eureka moments, and occur when a simple solution to a complex problem suddenly occurs to someone who has long pondered the matter, and if the many stories about them are true, they usually do so at unexpected times.

My tea was cold before I recovered. I was angry with myself that I had not worked out before then why differences between a pair of species in one country might not apply in another country.

At the same time, I was elated to have found a clean solution to the problems of taxonomy and biogeography that had bugged me since I first worked on corals in Western Australia. These weren't just my problems: the question 'What are species?' had been plaguing taxonomists, evolutionists and many others for as long as I could remember.

I called my concept reticulate evolution, having come across the term in a book called *Plant Speciation*, written by Verne Grant.[26] Grant's reticulate evolution and mine have plenty in common, but he coined the name to describe something that he thought was peripheral, whereas I believe it's at the hub of the organisation of Nature, and the main mechanism by which most groups of plants and animals change over time.

The basis of Darwinian evolution – natural selection and survival of the fittest – can be seen in action everywhere and is, I believe, as do all sane scientists, beyond question. However, Darwinian and reticulate evolution are poles apart on practically every point when it comes to when, where and how species come into existence.[27] Yet there is overwhelming evidence that both are correct: how can this possibly be?

The very short answer is that Darwinian evolution (and by that I mean neo-Darwinian evolution, the 'neo-' bit being genetics, which hadn't surfaced in Darwin's time) sits under the umbrella of reticulate evolution. It's like the icing on the cake, the cake being reticulate evolution.

The slightly longer answer, which focuses on how the term 'species' is understood, is that most interpretations of Darwinian evolution are based on the premise that species are reproductively isolated units (that is, they do not hybridise), whereas reticulate evolution is based on the premise that they are not. That initially seems rather facile, but the distinction is fundamental. We humans need units of some kind in order to name, describe or map – to do almost anything that involves communication.

If there were another life form able to communicate in terms of continua, our whole concept of how Nature is organised and changes over time would be very different. The crux of the matter is that Nature seldom forms units of any kind below the level of groups of species called syngameons (as I'll explain below); we must impose units, most commonly via species names, on natural continua because we have no choice in the matter, at least I haven't thought of one.

What, then, is a species? This simple question has dominated the thoughts of evolutionists for two hundred years.

Mind experiments can help explain the differences between Darwinian and reticulate evolution, 'evolution' referring, in both cases, to the process leading to the formation of a new entity that a taxonomist calls a species.

Experiment one: imagine what would happen if all the surface currents in the ocean stopped. Most fish, reptiles and mammals would still be able to move around because they can swim, but most invertebrate species would stay put because they depend on their larvae being moved by currents to disperse. Over a very large amount of time, these isolated animals would gradually evolve (by Darwinian natural selection, leading to survival of the fittest) into separate species, each endemic to the place where it lived. Now imagine the opposite, a situation where currents are so strong and variable that larvae of all descriptions are regularly dispersed everywhere. Under such conditions there would be few species because they would hybridise everywhere and those that existed would have huge distribution ranges.

While neither of these extremes has actually happened, our oceans have oscillated between them, through continental drift, sea level changes, and random changes in the paths and gyres of ocean surface currents. Changes in ocean currents and the larvae they carry mean changes in the pathways of gene flow, paths that are continually being broken and reforged in time and space.

In this scenario, evolutionary change is being created by currents, and currents don't care about the larvae they carry. This evolution has nothing to do with natural selection, competition, survival of the fittest, or any other biological process.

Experiment two: this one is on land. Today, botanists estimate that there are more than 850 species of what we call eucalypt trees growing in Australia. Imagine their fate if the continent gradually dried out. Over many generations, all would retreat towards the coastlines where they could find enough water to survive. Now imagine the opposite, a geological interval when rainfall is widespread across the continent. The eucalypts would spread out from the coast and gradually populate the whole continent.

These extremes never happened either, but the climate of Australia has oscillated between them for all known time. If eucalypt species were reproductively isolated, these climate changes would simply result in distribution changes, but most are not reproductively isolated, most can interbreed with many other eucalypts, so changes in rainfall patterns create changes in the pathways of gene flow among species, paths that are continually being altered in time and space. As with experiment one, this is evolution but it's evolution created by environmental changes rather than by natural selection.

Clearly it's not that simple; how could it be? I'm talking about how Nature is organised and how it changes, one of the most complex subjects imaginable.

Most evenings I can hear laughing kookaburras from Rivendell, beautiful birds that sound and look the same wherever they are, more or less. This is because they can move their genes around their home range – much of Australia – in sufficiently few generations to be able to maintain themselves as a single, genetically cohesive unit, and being a single cohesive unit they can evolve by natural

TIME AND PLACE

selection. From my study, I also see many poplar gums, eucalypts with distinctive leaves, bark and flowers. However, if I drive away from my house for an hour or so, up or down the coast, these trees start to look slightly different, and if I keep on driving they become very different, until, according to my field guide, they are not poplar gums at all, they are other sorts of gums. What, then, is the range of poplar gums? There isn't one, or at least not a well-defined one, because they are not units. What, then, *is* a poplar gum? They are something that was named from pieces of a particular tree growing in a particular place. Every other tree that carries this name is, ever so slightly, *not* a poplar gum. Poplar gums are units of human creation; they are not entities that actually exist in Nature.

So where do corals fit into all this? They vary a lot with local environment, so we need to revert to imagination again to remove that complication. Some corals are like laughing kookaburras in that they look the same wherever they occur, but most are like poplar gums, although with one huge difference. Most corals live in the Indo-Pacific Ocean, which encircles two-thirds of our planet. They would have to learn to fly if they were to move their genes around that much space to remain a cohesive unit. Yet corals, and many other groups of marine invertebrates, have tackled that problem very successfully – by producing larvae capable of long ocean journeys, not long enough to form a kookaburra-like unit, but long enough to retain some common identity from one country to the next, and so on across their whole distribution range.

Admittedly that sounds like cheating on my part: if I claim that a species known in Fiji also occurs in Madagascar, where it looks rather different, am I not just making two species into one? If at some future date someone tries to interbreed these two and finds they can't, am I not proven wrong? Well, yes and no. If that 'species' also occurs in Chagos, the Philippines and on the Great Barrier Reef, and is a single entity over every part of that range, we may have a continuum. Only where two of these entities can be seen to occur together in the same place (and so have overlapping ranges) do we have separate species.

Is all this not just old-fashioned geographic variation? It appears to be, until it is studied more closely. Unhappily for the student who does this, it just reveals a need for more study, for if the species is not a reproductively isolated unit, then the more it is studied, the deeper the problem gets. I will return to this below, but suffice to conclude here that there is no taxonomic solution to this very basic problem other than to stop imagining that all species are well-defined units.

One question that continually arises is: how prevalent is reticulate evolution? One way of estimating this is to find out which species are reproductively isolated (that is, form units), and which form interlinked groups. I turned to our 2000-odd domesticated plants and animals to try to find an answer and found that almost all can interbreed with other species: almost all are hybrids or could-be hybrids. That includes my dogs, which have a proven pathway of evolution that takes in wolves and even coyotes here and there. In fact I am sure it includes me, for in me the Neanderthals live on. My ancestors didn't kill them off, at least not all of them; they copulated with them, probably thousands of times, and sometimes the resulting progenies were better off as a result. It's a happy thought.

I first described reticulate evolution in *Corals in Space and Time*, the same book that initially revealed the existence of the Coral Triangle, as I will later relate.[28] It was published in 1995 by both the University of New South Wales and Cornell University and was reviewed that same year in an article in *Science*.[29] This book gave the Coral Triangle a flying start, but reticulate evolution remained in the doldrums for a long time, such is the enormous inertia of neo-Darwinism. The consequences of reticulate evolution, however, are far-reaching for almost every facet of biology, because evolution must be taken into account if the subject matter has anything to do with species. Only in recent years has it started to come to the attention of other areas of science, often being heralded as a new discovery or a new way of computing data. One day it will be seen for what it is: the central process in the mixing and remixing of genes along the pathways of evolutionary change for most plants and animals.

An important footnote to this subject is that Darwinian evolution is change for improvement – through natural selection – whereas reticulation is just change. The only improvement it might bring about is in the creation of a new genetic entity that Darwinian selection can then work on. The two diagrams below might help you follow this, provided you remember that they represent both time and space, the latter being a third dimension and too hard to draw.

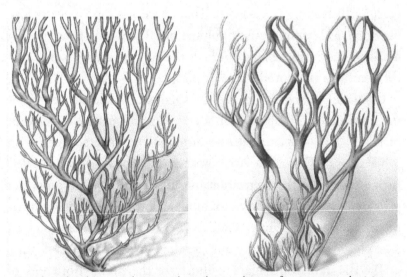

Hypothetical evolutionary change in identical species lineages from an ancestral origin to present time. Left: Darwinian evolution proceeds by divisions of clades through natural selection. Species are mostly defined units which have a time and place of origin. Extinctions occur by termination of lineages. Cladistics should replicate this phylogeny and indicate present-day affinities. Right: Reticulate evolution proceeds by both division and fusion of clades through environmental changes. Species are ill defined and have no time or place of origin. Extinctions occur by both termination and fusion of lineages. Cladistics will not replicate this phylogeny nor give a true representation of present-day affinities.

Another footnote is that Darwin (to whom I often talk, on account of being a little crazy) recognised the fuzziness I speak of and seems happy with reticulate evolution. In fact he thought about it himself, which is why he started gathering information about domesticated plants and animals all those years ago.

We now have computer programs that produce results from the otherwise unmanageable amounts of data that DNA studies produce, and similar programs to do our thinking for us when we analyse morphological data. The results of most types of computer analyses are unfused branches of a would-be evolutionary tree, because the programs cannot allow branches to fuse. Most programs that produce these diagrams, usually called cladograms, do so by a process called cladistics, the brainchild of the German biologist Willi Hennig (1913–1976). That process is supremely logical, provided, as Hennig himself pointed out, there is no gene transfer between the entities being studied: the process must assume that these entities are reproductively isolated.

Cladistics comes into its own on small scales, such as determining affinities of groups within a species – for example, between races of *Homo sapiens*, because humans cannot interbreed with other species. Cladistics also comes into its own on large scales, such as determining the relationships between the main groups of primates, as these taxa do not interbreed either. However, at intermediate scales, where interbreeding does occur, cladistics can give a false picture of evolutionary lineages that have occurred in Nature.

The diagrams on page 229 illustrate the point. Both sets of imaginary evolutionary pathways, if analysed using cladistics, might give a similar, or even the same, result. That result is true for Darwinian evolution but false for reticulate evolution at the level of species groups. If we call these groups genera, then cladistics may show an incorrect phylogeny.

I have little doubt that this matter will get no more than lip service for many years to come: why find fault with a nice cladogram produced by a computer that readers and editors alike will be happy with? The reality, however, is usually less satisfying; reticulate evolution predicts fuzzy geographic boundaries, fuzzy morphological distinctions, and fuzzy genetic distinctions due to multiple evolutionary pathways. It even says that binomial nomenclature is fuzzy, and that the synonymies that taxonomists love to play with may vary geographically. All very inconvenient, and often confusing, truths.

Curiously, the title of Darwin's ultra-famous book on the origin of species is almost a misnomer. It is an account of *why* new species evolve – because of natural selection and survival of the fittest – but it says nothing about the actual *mechanism* by which species originate. It's a bit like Newton's laws of gravity, a quantum leap in physics to be sure, but Newton had nothing to say about the mechanism by which gravity operates, something that initially eluded even Einstein.

Not surprisingly, thousands of evolutionists and biogeographers have jumped in to fill the void. In fact whole schools of thought, dominated by vicariance and phylogeography, have arisen, each giving birth to dozens of theories all based on units that can divide by Darwinian evolution, but not fuse.[30]

To highlight the point, there are about 3 metres of shelf space in my study holding just a fraction of the books on this subject. I have read most of them because I once set about writing yet another book on the subject myself. I was going to call it, surprise surprise, *Reticulate Evolution*, but it was never finished because climate change and the mass bleaching of corals raised their ugly heads, and that, for me, was a game-changer. However, when reading all those books and the hundreds of research papers that underpinned them, I saw that reticulate evolution overwhelmingly refutes one hypothesis of the mechanism of evolution after another. It also gives direct answers for most of the biogeographic, evolutionary and taxonomic problems that so many authors have pondered and complained about. In short, these authors would not have written what they did if they had embraced the significance of the differences between the two diagrams on page 229.

I will end this discourse by explaining my reference to syngameons. A syngameon is a reproductively isolated unit, the old definition of a species. Let's not split hairs over this: for most organisms, syngameons and species are not remotely the same things. For laughing kookaburras they probably are; for eucalypts and most other major taxa they're not. Depending on your reference book, eucalypts are now divided into two or more genera, with *Eucalyptus* divided into subgenera, which in turn are divided into species. Those species are divided into subspecies, and subspecies into varieties.

Are the poplar gums around Rivendell, you might therefore ask, a variety? No, they're not; they're part of a continuum that keeps on keeping on – forming finer branches long after all books throw in the towel. So where, in all this is, are syngameons, the slices of the pie that are indeed reproductively isolated? Perhaps roughly at the level of subgenera, for most species of eucalypt can readily be crossed with many other species, but certainly not all other species.

If this is so, why not draw the line, for eucalypts, at the level of the syngameon (perhaps a dozen big units and many laughing-kookaburra-like small ones) and leave it at that? Many reasons. For starters, the taxonomic information loss would be enormous; syngameons are very likely to have no defining morphological characters; syngameons cannot be identified from DNA using cladistics; syngameons *can* be identified by thousands of cross-breeding trials, although the results will vary geographically.

All this raises the question of how, when and where species

originate. Looking at those two diagrams again, note that Darwinian evolution has species forming at specific times (and places), while reticulate evolution has species forming without a time or a place of origin. Only syngameons have that time and place, vague though these might be. Note also that extinctions may be similar in both cases; however, with reticulate evolution what appears to be an extinction may be a genetic merger created by an environmental change. The latter can be seen in the fossil record (as 'punctuated equilibria') and in virtually any aspect of present-day biology.

So what, in the natural world, can one make of all this? Not much, except to be aware of the fuzziness and not assume that a population in one part of the world has the same properties as a population of the same species in another part. When it comes to conservation, that's a lot of food for thought.

Isn't reticulate evolution also a good excuse to let taxonomists off the hook when they're not too sure about their species? Certainly, but only for those who don't delve into details or work over big geographic ranges.

For me, as for so many others, the mechanisms that change the life of our planet is an all-engrossing subject, one I would dearly love to write about, but the need to do my bit to help keep that life alive has become much more important.

Origins of The Reef

Darwin's theory of evolution was not that great man's only profound and everlasting contribution to science. His hypothesis of how atolls form is infinitely less deep and certainly less well known, but for me it was the theory that most mattered when I first started working on coral reefs. My interest was triggered during the Stoddart expedition, when we first dived down the outer face of Tijou Reef and found that it is also both the outer face of the Great Barrier Reef and the western edge of the Queensland Trough.

My journey with this subject has the longest time span of any scientific concept in my life. I first read, and wondered, about the origin of Australia itself when I was a dream-laden kid at Barker in 1960. Then I brought the Great Barrier Reef into that scenario when writing *A Reef in Time* in France nearly fifty years later.[31] To put all this into context I now need to retrace Australia's journey from its motherland, the supercontinent of Gondwana, where Antarctica still resides.

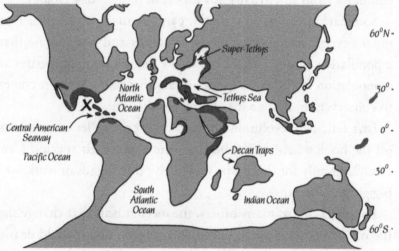

Continental positions during the Late Cretaceous, about 80 million years ago. Surviving reef limestone is shown in dark grey. X marks the spot where a giant asteroid hit the earth 65 million years ago.

By the Early Cretaceous (around 130 million years ago), Africa, Madagascar and India had rifted free of Gondwana, and India was on its dash to the equator. At that time Australia was dawdling, so that by the Late Cretaceous (about 80 million years ago) it was only just setting out. However, the new continent carried with it all manner of bizarre plants and animals, including *Glossopteris* and other temperate rainforest trees that we still have, and a suite of giant reptiles, including many dinosaurs, all tolerating the seemingly lethal seasonal daylight regime that Antarctica has today. When it finally broke free, Australia was isolated in all directions, allowing its marsupials to evolve free of competition from the more advanced mammals of other continents. At least that was the stage-set for the evolution of Australia's unique terrestrial life that we have long been familiar with. The story of its marine life is another matter, one that only corals can tell.

Sixty-five million years ago, a massive asteroid hit what is now the Gulf of Mexico, shaking the whole of Earth and triggering the last great mass extinction, which wiped out the remaining dinosaurs and much else besides. About that time, Papua New Guinea started rotating counterclockwise away from northeast Australia, creating the Queensland Trough and, I have to suppose, the shallow platform on which the Great Barrier Reef now resides. Reefs did not proliferate anywhere at this time because major carbon dioxide peaks snuffed most of them out, worldwide.

The Antarctic Circumpolar Current was established about 40 million years ago when South America rifted free of Antarctica, the last continent to do so. This formed a marine thermal barrier around Antarctica, helping the polar ice cap to form and giving most of Australia its all-time lowest water temperatures, while the continent was less than halfway along its journey to where it is today. At that time, reefs would not have formed anywhere around Australia as we now know it. However, they did form along the north coast of New Guinea, for Greater Australia (meaning Australia and New Guinea combined, as they have been for most of their existence) had come within seeding range of the coral reefs of Asia.

By the Early Miocene (around 20 million years ago) terrestrial Australian life (seabirds presumably excepted) was still isolated from the rest of the world, but Australia's marine life wasn't. Minimum temperatures were often higher than now, and we know that coral reefs proliferated down the coast of Western Australia as far south as Perth, so there's no reason whatsoever to believe that the same did not happen down the east coast.

All this points to the Great Barrier Reef, roughly the size and shape it is now but probably bigger in the south, being formed well over 20 million years ago.

On 12 April 1836, Darwin knew nothing about continental drift – not even he imagined that – but after climbing a mountain slope in Tahiti, he gazed out on the beautiful island of Moorea, often called the Bali Hai of the South Pacific, and that night he made an extraordinary entry in his journal: '. . . if we imagine such an island arc after successive intervals, to subside a few feet in a manner similar to but with a movement opposite to the continent of South America; the coral would be continued upwards, rising from the foundation of the encircling reef. In time, the central land would sink beneath the level of the sea and disappear, but the coral would have completed its circular wall. Should we not then have a lagoon island [atoll]? Under this view, we must look at a lagoon island as a monument raised by myriads of tiny architects to mark the spot where a former land lies buried in the depths of the lagoon.'[32]

He then turned this wonderful piece of deductive logic into his unlikely theory of the origin of atolls. As the greatest geologist of his time, Sir Charles Lyell, put it: 'I am very full of Darwin's theory of Coral Islands [atolls] . . . Let any mountain be submerged gradually . . . there will be a ring of coral and finally only a lagoon in the center. Why? For the same reason that a barrier reef of coral grows along certain coasts: Australia etc. Coral islands are the last efforts of drowning continents to lift their heads above water.'[33]

Darwin developed this theory just by looking and thinking, nothing more. Another of his eureka moments? Certainly looking at atolls can prompt pondering – they encourage any amount of that – but seen from the sea, atolls keep their secrets. It's only when seen from above, from an aircraft – in short supply in

Darwin's time – or on a chart, that they are seen for what they are, which is usually a string of narrow reefs that form a circle or an irregular loop, some dotted with little cays. The centre of the loop, the atoll lagoon, is almost always shallow and full of sand, while the outer wall plunges to a great depth.

Darwin's theory conjures up a spectre of mountains conveniently sinking at just the right pace for corals to keep growing, yet this must rarely be the case. Some mountains go in the opposite direction – up – and these often have dead reefs on their slopes. Others sink too quickly, and if these have reef remnants at all they are dead for want of energy-giving sunlight. Atolls only form when the rate of sinking is tolerable for coral growth, and this must happen in the context of ever-changing sea levels. So the chances of an atoll forming are really quite slim.

Was Darwin right? About twenty of his contemporaries got stuck into the arguments. If he was right, reefs would be thick, in places very thick. If he was wrong, they would just be veneers atop the submerged remains of mountains. Most of the conjecture was about how reefs would survive changes in sea level resulting from polar ice sheets forming and melting during ice age cycles, creating sea level changes of 100 metres or more. Reefs would be shaved off when the sea level went down and then would need to be rebuilt as it rose. There were many views published about these theories and variations of them, and they clearly nagged at Darwin. Just before he died, although rather more preoccupied with human evolution than with reefs, he wrote to one of his antagonists, Alexander Agassiz of the Museum of Comparative Zoology, Harvard University: 'If I am wrong, the sooner I am knocked on the head the better . . . I wish that some doubly rich millionaire would take it into his head to have borings made in some of the Pacific and Indian atolls.'

As is well known, this happened some fifty-six years later, when the United States Atomic Energy Commission, in preparation for nuclear testing, sank a series of deep boreholes in the northern Marshall Islands. There was immense public interest in the issue: Darwin had predicted that atolls might be 5000 feet (1525 metres) thick. Two boreholes reached volcanic foundations at 1267 and 1045 metres. Darwin was right, but how could he have been that accurate?[34] I have no idea, and nor have the dozen or so reef geologists and historians I have talked to about it.

Before all this happened, and to supposedly prove or disprove Darwin's hypothesis, some crazies from the University of Queensland decided to take a core from the Great Barrier Reef, apparently not having observed that Australia is just about the oldest continent on Earth and would be exceedingly unlikely to be sinking, rising, or doing any other such thing. In 1937 they drilled a hole on Heron Island, and when that didn't work out, because their equipment wasn't the best, they drilled several more holes in and around the far northern ribbon reefs, and when those didn't work out either they gave up. In hindsight this was fortunate, because there was a lot of interest in these drillings, and newspapers of the time would surely have pronounced Darwin wrong, in line with the narrow-minded, Christian-dominated view most people had about his theory of evolution.

In the 1970s, a group of geologists again began taking deep cores from the Great Barrier Reef, in what developed into the most expensive research project ever undertaken on it. This coring (the word 'drilling' being by then taboo because of its association, at least in the minds of the permit peddlers in the Great Barrier Reef Marine Park Authority, with oil exploration) to determine the age of The Reef became famous, especially as its leader, Peter Davies from the Bureau of Mineral Resources, gave engaging talks on the subject at several reef conferences. Peter's work resulted in two basic notions about the Great Barrier Reef – firstly that it is (horizontally) wedge-shaped, being thicker in the north because it is older in the north, and secondly that it is very young as reefs go. As far as being wedge-shaped goes, the Great Barrier Reef is as thick as the distance between the bedrock and the ocean surface, something that has nothing to do with age. And as for being a baby, this view came to a much dramatised climax in 2001, when an international drilling consortium using a 50-tonne drill to go coring announced that the central Great Barrier Reef had started growing about 600 000 years ago, a conclusion endorsed by a football team of reef geologists.[35]

I countered this conclusion in *A Reef in Time* seven years later by pointing out that it's unreasonable to suppose that just because old reef (at the bottom of their boreholes) isn't there now it never was there. Older reefs may have grown and been eroded away many times following sea level changes. I'm still waiting for a response to that, and am fairly sure I won't get one because I don't believe there is one. If extensive reef development occurred on the west coast, why would reefs not have formed in the east, when environmental conditions and bathymetry were at least as favourable?

The ability of coral larvae to make long-distance journeys, which underpins what I have just described, was only a dawning notion thirty years ago and would have been unappreciated by geologists then, some of whom expressed the view that corals were dispersed by seafloor spreading – by hitching rides on islands and continents as they moved about.

This geocentric view of the origins of reefs is now being clarified by mapping the surface of reefs in detail, using high-resolution, three-dimensional bathymetry.[36] Today we have confirmation that the ribbon reefs had a deeper-water forerunner from times of lower sea level, something I've seen many times when diving, the second line of reef being clearly visible at depths of 50 metres or more. However, I wonder if we'll ever know how many times this has happened. Very likely we never will, as neither drilling into reefs nor mapping their bathymetry will give an answer.

And given the amount of time involved, the story is certain to be anything but straightforward.

Of course Darwin's insight into the origin of atolls and my view about the age of the Great Barrier Reef could have been arrived at by getting together information from all relevant fields and just integrating it. However, this bit of integrating occupied me for thirty-five years. One mustn't rush these things.

Darwin's theory of atoll formation, where a fringing reef grows around a subsiding mountain (left), which ultimately becomes an atoll after the mountain has become completely submerged (right).

The Coral Triangle

When John Wells gave me his wall chart of coral distributions, back in the autumn of 1975 on my first trip to America, I wasn't too keen on copying it all out, so it sat on a shelf in my study gathering dust until fate intervened in the form of a magical machine I bought from Woolworths: a Commodore 64 computer. This was mostly designed for playing games, but with two fingers to the task and a little help from a friend who'd used a computer, I soon had John's table typed in and, lo and behold, I could easily make changes without reverting to scissors or sticky tape. Better still, I could print out lists of countries where any coral genus had been recorded, as well as lists of genera in each of those countries. My study at Rivendell became littered with piles of printer paper, excellent fodder for a new contour map of the world's coral genera. I reported on this to the Fifth International Coral Reef Symposium in Tahiti, in 1985.[37]

In the meantime, I set about putting these records on AIMS's new computer, an IBM mainframe monster that filled two large rooms. At first I'd regarded computers with suspicion, mostly because the electron microscope I used for my PhD, another monster, had taken a lot of time to master and I'd never used anything like it since. The same sort of thing seemed to apply to computers; better to leave such jobs to a technician.

Using the new IBM was a big mistake on several fronts, the first being that I only narrowly escaped that most terrible of fates – being drafted onto a committee.

'Charlie, I gather that you're one of the main users of our computer, so I want you to form a computer users' committee and report on what we should be doing with it,' said John Bunt, the director.

I'm not forming any bloody committee.

'Sure, John, love to. We've got one of those new electric ones, haven't we?'

That was close. Doesn't he know the computer is only used for playing chess?

About this time, I took my Commodore to AIMS for some driving lessons. The manager of the computer centre was aghast.

'Well,' he said, glaring at me, 'if you want to go it alone, don't come to us for help.'

So I didn't. I took my computer back home and started entering species records onto a spreadsheet, which I could figure out myself. I called the file Coral Geographic and that was the start of what was to become a very big headache, for distributions depended on taxonomy, and coral taxonomy at that time had a long way to go. The one saving grace was that contour maps of species richness don't need names, and so as I worked through the many complexities of my book *Corals in Space and Time*, I decided to bite the bullet and include a contour map of the global diversity of corals at species level, a much more complex undertaking than maps at generic level. This was beyond the capability of my computer, not to mention my ever-doubtful skill to use it, so I traded it in for an Apple and with Mary's help (well, to be honest, she just did it) eventually produced a map generated by a mixture of a spatial data program and some numerical fiddles. This compilation clearly showed something I had long known but apparently nobody else had – it was the Indonesia-Philippines archipelago which had the world's greatest diversity of coral, not the Great Barrier Reef.

This finding created a good deal of consternation among conservation agencies because the Great Barrier Reef, by then given World Heritage status and managed by a big Australian government agency, had long been assumed to offer permanent protection for the world's centre of reef diversity. The reality, as shown by my map, was that few of the world's mega-diverse reefs occurred within any well-managed marine park or had any legal protection. Worse, these reefs, which at the time I believed to be confined to the Indonesia-Philippines archipelago, were in a region where human population densities were high by most world standards, and so was the environmental impact of all these people. Such concerns soon precipitated one of the biggest information quests in the history of marine biology – to delineate the global centre of coral diversity, the boundary enclosing a region with at least five hundred species.

The name Coral Triangle came after a bottle or two of wine on the back deck of a yacht, which had been charted by Conservation International for a group of us to survey the corals and fish of Milne Bay, Papua New Guinea. In my experience, bottles of wine often have something to do with such developments.

The questions that naturally arose were: where exactly is the centre of coral diversity, and how distinctive is it? These were simple questions but they took a dozen expeditions, mostly run by American conservation agencies, to answer. I went on most of them, working with other field-going coral specialists as time and opportunity allowed. The shape of the Coral Triangle kept changing, especially when we extended it to the east to include northern and eastern Papua New Guinea. In the end that left only one major unknown – the Solomon Islands. To tackle this, the Nature Conservancy chartered the *FeBrina* from Papua New Guinea, an excellent boat that we joined in Honiara in 2004. For the first time in years I felt I was back at sea doing what I used to do on beautiful undamaged reefs.

I digress. When we'd finished delineating it, the Coral Triangle had an area of 5.7 million square kilometres, but it wasn't actually a triangle at all; it really had no particular shape, but no matter, the name stuck.[38]

The political response to our findings was astonishingly prompt. In September 2007, twenty-one world leaders attending the Asia-Pacific Economic Cooperation summit in Sydney recommended the establishment of the Coral Triangle Initiative, for the protection of coastal and marine life in the region. The move was endorsed by the Indonesian government and launched at the United Nations Framework Convention on Climate Change, Bali, in December of that year. That was an extraordinary event, with the delegates a mix of politicians, conservationists and religious leaders from the seven countries in the Coral Triangle region. I was given a central role but was told I had ten minutes to say my bit and was advised to leave science out of it. Indeed, it does seem that prayers more than science carried the day. Afterwards (when to my relief, a little booze had been furtively broken out) I kept repeating that I had no funding to carry on with this work, and that the information which had led to the identification of the Coral Triangle had never been published, and couldn't be other than on a website, there being so much of it. That brought another prompt reaction. The US State Department and some conservation organisations promised money for a website. Then the Global Financial Crisis hit: goodbye website. The mapping went on hold, but there was much else to do, especially in taxonomy.

It has been curious to watch the development of the Coral Triangle Initiative and the science that surrounds it. The Coral Triangle is now the subject of excellent books as well as television documentaries, and scientific articles about it number in the hundreds. Full marks to the conservation agencies that played such an essential role in this development, but after that job was done they turned to management rather than exploration. I waved farewell at that point and am content to have done so. Our very big website soon took back the reins and now the Coral Triangle includes another country, Brunei, much to the consternation of the Coral Triangle minders, I'm sure.

Sorting corals on the back deck of the *FeBrina* at the Solomon Islands.

Big Pictures

A boy and a book

In 1998, with Katie principally living with Kirsty, and a friend offering to look after Eviie at Rivendell, Mary and I were able to organise a working holiday on Norfolk Island. From the late-eighteenth to the mid-nineteenth century, Norfolk Island had been a penal colony, among the worst of Britain's notorious colonial prisons and a place from which suicide was the only escape. The history of that time is still everywhere to be seen, especially the old prison with its grand gates leading to hell: such a contrast with the peaceful natural beauty of the place.

Because of its isolation I knew that the corals there were likely to be a depauperate but interesting lot, and they were, forming communities with a species mix not found anywhere else. There was next to no reef, but with clear water and nothing to disturb them, the corals grew in profusion in a shallow bay protected from the waves that hammer most of the coast. I had already named a new species from the island, *Goniopora norfolkensis*, collected for me decades earlier. On this trip I took what time I had to work out what species were present, but it seemed that several more might be found with further work, something I still hope to do.

Mary could only snorkel with me, for a very welcome reason: she was pregnant. Our son Martin was born in June of that year, a beautiful baby boy born without problems. Could it be that I was leading a normal life? Kirsty was always happy for us, but of course I wondered how she must have felt with so much going smoothly for Mary when it hadn't for her.

By the time *Hermatypic Corals of Japan* was published (1995), a single, broadly cohesive global taxonomy of corals was becoming reasonably well established. Most old species based on single specimens had been amalgamated, and many new ones had been discovered, mainly from the vast amount of field work my colleagues and I had undertaken. However, the amount of unpublished work we had completed after research in so many countries was getting out of hand. I felt there was little point in publishing accounts of the corals of each country; what was needed was a *Corals of the World*.

Scoping a book of this size turned out to be a discouraging business; the job just looked too big, too complicated, and needed too many photographs. There were plenty of publishers willing to take it on, but the price they'd have to sell it for would put it beyond the reach of most students, certainly those in developing countries. I didn't want to write a book just for rich people, I wanted it to be for those who would use it.

At this gloomy point we got an unexpected helping hand from John Jackson, who'd organised my trip to Clipperton Atoll. John was the owner of Odyssey Publishing, a small company based in San Diego, and he used to come to Australia about once a year to catch up with his shell-collecting buddies in Perth. He always stopped off at Rivendell on the way home.

'You know, Charlie, Mary could build this book,' he said as I was taking him to the airport for his homeward journey.

'Don't think so, John. She's never so much as made a postcard.'

'No harm in asking.'

John didn't know the first thing about computers, in fact I had only recently persuaded him to get one, but he did know a lot about people.

Mary, who was already working on the science, said she would think about it. Then she bought a copy of QuarkXPress, a publishing program which at that time was a beginner's nightmare. A week later she was building page designs.

I managed to convince Russ Reichelt, then director of AIMS, that the project was doable. The AIMS council thought so too, and agreed to pay the printing cost – more than half a million dollars – on condition that sales recouped it. That cost made it one of the most expensive printing jobs in Australia's history. Another condition was that AIMS's logo be on the spine as 'publisher'; it was a millennial celebration project that would impress Canberra and send an up-yours to the CSIRO, who were trying to gobble AIMS up. Building the book would be expensive but fortunately Mary and I had obtained a grant to build an electronic key to identify corals, and that had much the same content, especially photographs, as a book would have. Well, sort of.

Mary bought a new, high-end computer for the job, with a top-of-the-range hard drive – of 19 gigabytes. I had thousands of photos, taken during my travels, but these weren't nearly enough. Many people offered to help, and did so generously. There being no digital cameras in those days, all the photos were 35-millimetre transparencies and all had to be scanned, corrected, and sometimes scanned again. In the end we had 70 gigabytes of photos, which had to be stored in a bank of hard drives and laboriously backed up on tape. Every hard drive eventually failed and had to be replaced, but the back-up process Mary strictly followed always came to the rescue. Today, photographers return from trips with thousands of digital photos, all checked and sorted. I used to return with rolls of Kodachrome, never quite knowing what was on them.

In late 1999 our small team – Mary and me, my assistant Laura Carolan and our can-do friend Gary Williams – finally popped the cork. The bottle had waited years for this moment, years of solid work that were hectic, often frustrating but always rewarding. It was a good team.

I took our book from Rivendell to the printer in Melbourne on our tape drive, still having no publisher for it. The coral bible, as it came to be called by others, was launched at the Ninth International Coral Reef Symposium in Bali amid a fair measure of applause.[39] Later it attracted some awards, including the Darwin Medal at the following Coral Reef Symposium.[40] Although the book was sold to students at not much more than the cost price, it recovered AIMS's investment and went on to make a modest profit. That's amazing considering I was its financial manager, and hadn't even thought of such things as the cost of air-freighting a three-volume book that weighed 7.5 kilograms. John Jackson helped there, becoming our agent for most overseas sales.

The book went on to be a sellout, and would have been reprinted by a commercial publisher, but as I will relate, we had other plans: to replace it with a website of the same name. In the meantime, Mary and I followed it up with our long-planned electronic key to coral species called *Coral ID*. This was published on CD-ROM discs, now an outdated technology, but at the time it was much appreciated as it cost users little and enabled them to identify corals quickly and surely without having to trawl through endless printed pages. As with the book, this publication is now part of our *Corals of the World* website and is free to all users the world over, which was what we'd always wanted.

Paths of conscience

I had imagined, wrongly as it turned out, that publishing *Corals of the World* would be some sort of culmination of my work on coral taxonomy. It did serve to create a world view of corals, something I'd worked towards for a very long time, but I soon discovered that answering a lot of questions only begs the asking of a lot more questions. And so the book turned out to be more of a beginning than an end.

The new millennium was also the beginning of a different relationship with AIMS. It started out well enough, with the appointment of a new director, but after a couple of years I felt that my time was being wasted. I decided to quit the executive, the internal governing body of the institute, a job I'd been embroiled in for more than a decade. I remained a troublemaker of course, I had no choice: the staff had good reason to be concerned about the way the institute was being run, including the composition of the council, none of whom knew anything about marine science.

Perhaps in retaliation, but actually to my delight, the director announced at a summit staff meeting that *Coral Geographic*, my long-standing mapping program, would not be accepted as an AIMS project. Somebody, not me, had proposed it should be, but it was my personal hobby and had been for thirty years. According to the director, I had no projects at all. The timing was perfect, so chuckling to myself I walked out of the meeting, up to my office, and phoned Mary.

We had been pondering the possibility of an extended stay in France. Mary had seen a lot of Europe as a child but I had never left Australia until I joined AIMS. Eviie and Martin had both travelled a lot, but only as sightseers or to visit Mary's family in England; now they were of an age where they could absorb the language and culture of another country. Wanting to avoid cities, we looked for a place to live where the kids could go to small rural schools and get immersed in all things French. Katie, too, would come for part of the time. She had taken a particular interest in French and was keen to practise it.

After much hunting, Mary found the perfect house for us: an ancient, rundown but incredibly picturesque olive mill in a remote mountain region in the south. The mill, made of thick stone walls with a gabled slate roof, was perched on the side of a steep mountain slope, with a broad river at its feet and small streams on either side. I calculated that I could be away from AIMS for a year and a half if I combined my long-service leave with half-time work. Just what that work was I never specified but I reasoned that the director would approve anything to be rid of me. He did. After placing Rivendell and all our animals in caring hands, we left for France in July 2003.

Mary's French improved rapidly but I didn't give myself much of a chance. Most days, after driving the kids to school, I would spend a little time trying to chat to a villager and then seclude myself within the thick stone walls of a room at the bottom of our olive mill. With its own fireplace it was a cosy retreat that I turned into a study, complete with a very large crate of literature. The reading and writing I planned to do was much more important than learning a language.

For more than a decade I'd been worrying about the future of coral reefs, knowing that they're as vulnerable to climate change as anything on our planet. There were dozens of theories and counter-theories surrounding this subject. Points of view were bandied about, some in scientific journals but most in the popular media, where so-called sceptics were having a field day. What was the *real* truth? I wanted to work this out in detail, and eventually did.

My first job was to undertake a vast amount of reading in every field of science that seemed relevant. As the months went by I read far more than I ever had for any university degree. I can't say I enjoyed doing this at first; it reminded me of my PhD, where

I seemed to be going ever deeper into a series of unconnected subjects. But gradually, like peeling the proverbial onion, I began to see what the cores of these subjects consisted of, how they were linked, and what this meant for the future.

On our return to Australia my head was full to bursting with climate change issues. Unlike in France, the media in Australia was still giving prime time to climate change sceptics, and now that I could evaluate their worth with certainty they made me angry. Foremost among the sceptics regularly appearing on television was a professor of geology from James Cook University, whom I'd known for decades. It was easy for me to debunk his views and I did so with careful precision, one after the other, when we met. To no avail. He was back on television at the next opportunity, wheeling out the same pseudo-science and undermining the efforts of so many scientists who knew their subject and were giving their time to help the public understand it. Nevertheless, the professor did me a good turn by showing me that I could no longer put my head in the sand. Throughout my time at AIMS, I had appeared in many science documentaries but had always avoided contentious public issues if I could. That now changed.

Before we'd left for France, the Howard government had issued instructions that Commonwealth agencies were to have nothing to do with climate change. When we returned, AIMS had another new director, who was not one to buck the system. He ordered me not to speak to the media about climate change, whereas I was determined to do so at any and every opportunity. I had much greater security than my colleagues, thanks to a council decision to make my tenure pretty much unbreakable, but I was unhappy that other senior scientists seemed to have done nothing to prevent bureaucrats taking over and running AIMS their way. That made a joke of my role as chief scientist, so I quit that in much the same style as I'd quit the executive a couple of years earlier. I also decided to leave AIMS altogether, but at a time of my choosing.

When I did finally walk out of AIMS's front door for the last time, on a sunny afternoon in July 2007, I had cause to ponder. I'd been there when every other staff member first arrived. Throughout most of that time AIMS had been good at send-offs, giving people the opportunity to say things they thought should be said and spreading a little warmth around. Most departing staff were given a pretty piece of coral in a perspex box with a brass label; it had become a tradition. Thus armed, I had given farewell speeches on behalf of AIMS to about half its departing chairmen, most of its directors, and I don't know how many other staff of all descriptions. It was something I enjoyed.

The day I left, I waved to Jim the gardener on his tractor, got into my car and drove off without so much as a farewell cup of tea. I missed the opportunity to thank all those who'd helped me over the years, especially the support staff, the people behind the scenes. They never seemed to get a mention when someone else gave a speech, yet they were the backbone of the institute as far as I was concerned.

I left because bureaucrats had won and AIMS had become just a building. My ignominious departure was not a personal rejection. Most of the old guard departed around the same time, and with a similar lack of grace. Some were more or less fired, others were asked to stay on and declined. For my part, I had no intention of working in a place where bean counters and an electronic key to doors controlled my life.

The saddest part was that although the more experienced scientists learned how to put up with it, or found another job, beginners seemed to think it was normal. Normal it might have become; necessary or beneficial it certainly wasn't. I can't think of a better way of killing off the creativity of a creative person than telling them what to work on, where to work, and when to work.

People who do their job well need to be left to get on with it.

I pause here to emphasise this because of its importance in today's wider world. At a council meeting in 1995 the chairman, Ray Steedman, announced that he wanted me to produce a book about the institute's first twenty-five years. 'Warts and all,' he said. *AIMS: The First Twenty-five Years* was published in 1998, with historian Peter Bell the principal author and me the editor. In the introduction I wrote: 'AIMS is what it is because of close links among the people who work here. AIMS is isolated and has been a frontier organisation in many ways. It has always been full of prima donnas, factions, arguments, workaholics, sloppy dress, self-defensiveness, general infighting and non-hierarchical cooperation. Visitors in suits come from afar – mostly Canberra – and frown on all this and mutter about 'responsibility' and 'accountability'. To which the staff can justly reply that our culture, despite their efforts, is alive and well. And at the end of the day, where else can you find a track record of so much achievement that is real? We have long been, and still are, doing it our way – and curiously enough this tends to keep us ahead of the game.'

Less than a decade on, that atmosphere had gone and with it the creative zeal of the institute. It had been the place for me for a long time, but it was no longer. On my return from France the place had the air of a prison, with the bars of bureaucracy at every door.

Reefs in time

The interlinking of many fields of science that I began in France continued on my return to Australia, finally culminating in *A Reef in Time*, published in 2008. This book delved into how and why reefs have changed over all time scales and what this predicts for the future. A big part of that delving concerned reef environments.

When you think about it, coral reefs are extraordinary. They have evolved to live and thrive at a perpetually changing interface of land, sea and air. Compared to the relative constancy of fully terrestrial or fully marine habitats, the coastal fringe is exceptionally forbidding, and yet over hundreds of millions of years corals have made it their own. Their success can be attributed to the control they impose on their environment courtesy of their tight symbiosis with algae. This relationship provides the energy needed to build their three-dimensional matrices of stone 'trees' that are homes for the herbivores that keep seaweed in check.

There is a downside, though.

To cope with the physical ravages of their environment, and to keep their symbioses in order, corals live on a knife edge. They and the reefs they build go through never-ending boom-and-bust cycles. The booms occur when the sea level is mostly constant and atmospheric carbon dioxide levels are low; the bust conditions are the opposite. The role of sea level change is rather obvious: reefs are left high and dry when the level falls and they can die from lack of sunlight when it rises; it's a matter of catch up or die in the latter. (I will return to this subject below.)

Carbon dioxide, a gas rarer than argon, is critical for life on Earth. It is the essential ingredient for photosynthesis in all green plants and it's also essential for keeping our planet warm: if we didn't have it our oceans would freeze over and life as we know it could not exist. The mechanism involved has been known for two centuries: carbon dioxide is transparent to the short wavelengths of sunlight and so lets the warmth of the sun in during the day but acts to block this warmth from being re-radiated out at night, as that involves long wavelengths. Moreover, carbon dioxide is the fast-acting currency of the carbon cycle, transferring carbon between living creatures, rocks and air, and around again. All this makes it the controller of a strange mixture of life-giving properties of our planet, regulating Earth's temperature and being the major player in the chemistry of Nature's cradle, the oceans.

This is the background I painted before turning to mass bleaching, one of the two target syntheses of the book.

Mass bleaching is now a permanent fixture of the ecology of all coral reefs. At first it was restricted to El Niño years – the natural weather cycles that usually come every four to seven years and bring abnormally high temperatures to reefs throughout the tropical and subtropical world. It wasn't until 1998 that coral biologist Ove Hoegh-Guldberg put forward a hypothesis about how mass bleaching occurred, one that some scientists thought implausible when he first came out with it.

For hundreds of thousands of years, corals and their algae have lived together in symbiotic harmony. What Ove discovered is that if corals are subjected to too much temperature *and* light, their algae go into overdrive, producing too much oxygen, some of which remains as free radicals that damage their host cells. But corals can counter this by controlling the number of algae in their tissues. Most biologists believe they do this by expelling the algae, but it's more likely that they slough off the affected cell layer and replace it with a new layer that has no algae. Either way, the problem is that zooxanthellate corals cannot live without their zooxanthellae, and if they get free of all of them they die. It's all rather suicidal, and now corals are suffering the consequences, because they live in a world where temperatures increasingly peak above their precisely evolved limits of tolerance.

Not all mass bleaching is lethal. Sometimes enough algae remain to allow the coral to recuperate, but if no algae remain, death usually follows within months. Seeing entire coral communities turn white is a gut-wrenching sight, especially as it's not just the corals that die, but also most other animals in the community created by the coral, paving the way for a seaweed or slime takeover.

With the reality of reefs dying en masse, I wanted to estimate the possible flow-on effects this would have for all marine life. This was prompted by the geological record, which suggested that the demise of reefs gives early warning of a global environmental collapse, even a mass extinction. Evidence for this comes from several quarters, including the links between carbon dioxide and the carbonates that reefs are made of, and the sensitivity of corals to ocean carbonate chemistry. I tried, unsuccessfully in the end, to rope others into helping me to estimate how much marine life would be affected by a complete collapse of coral reefs, but the job was too big and there was not, and still isn't, enough scientific knowledge about our marine life to draw on. Taking this subject as far as we could, I was persuaded that about a third of all marine animals are dependent on reefs during at least one stage of their life cycle. If this is true, then the collapse of the world's reefs would indeed trigger an ecological collapse of our oceans. The geological record, full of suppositions though it is, clearly shows that this has happened many times during our own era (the past 65 million years), and I now felt I had a good case for proposing that we are creating conditions for a re-run. One of many re-runs though it would be, the next will be different because we humans will be among the victims. Will we be prepared? Of course we won't.

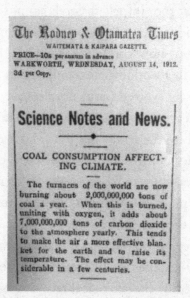

Coal, the stuff that once powered the Industrial Revolution, is now powering climate change. This prophetic excerpt from a 1912 New Zealand newspaper speaks volumes, yet we are still mining coal with reckless abandon.

I turned to sea level rise. As I continued with my studies this made a welcome change from being so doomy and gloomy. It had been proposed in many quarters that sea level rise would destroy our reefs, but those who argue this should look at the rate of rise after the last ice age (up to 1.5 metres per century). That was much faster than is likely to occur in our modern world of high sea levels, short of a melting Antarctica, and reefs coped with that rate well enough. What sea level rise *will* do is lead to the shifting or removal of the colourful zonation patterns most reefs now have. However, seen in the context of geological time, these patterns are abnormal, brought about by the unusually stable sea level we have enjoyed for the past ten thousand years or so.

The final target subject of *A Reef in Time* was ocean acidification, which is not an aspect of climate change although it has the same cause: carbon dioxide. Of all the syntheses I made in that book, those delving into the geological past were the most challenging because I constantly found myself in uncharted territory. But I could not let the subject go; it's critically important.

It's quite a climb to the very top of the highest Devonian reef of the Canning Basin, near the coast of north-western Australia, because the limestone, looking like a giant cheese grater, has a karst surface carved into ridges so sharp that even a goanna would be well advised to tiptoe carefully. This reef has a spiritual feel about it. It's so old that its years number as many as there are metres to the moon, yet 360 million years ago it was decimated, along with every other reef in the world, and they all stayed that way, home only to microbes, for more millions of years. What could have caused that? Ocean acidification and associated oxygen depletion (anoxia) from carbon dioxide could. Certainly, not all ocean catastrophes in the ancient world can be blamed on carbon dioxide, but in most cases I have searched in vain for alternative explanations, and I don't know why the role of carbon dioxide wasn't apparent to geologists long ago. The geological past has shown us many times over that the Earth has enormous reserves of carbon that can be converted into carbon dioxide and its chemical associates via a multitude of biological and geological pathways.

The chemistry of carbon dioxide when dissolved in seawater is generally understood and has been for a long time, although it is not straightforward because it is greatly affected by its concentration in the atmosphere as well as the temperature of the ocean. Like the gas in a bottle of soda water taken from a fridge and left in the sun, it does not stay dissolved when warmed. However, when it is dissolved it forms carbonic acid and that's the problem. Our oceans can never actually become acid – with a pH of less than 7 – there's not remotely enough acid on Earth to do that, but carbonic acid does make the surface of the ocean less alkaline, which increases the solubility of calcium carbonate, especially a form of it called aragonite, the stuff coral skeletons are made of.[41]

Corals are good at countering this, but as seawater gets less alkaline – meaning less super-saturated with carbonate, even slightly less – coral larvae will struggle to lay down the fine cementing layer that they use to attach themselves to rock, and if they do start growing, their skeletons will become brittle and then may cease to grow altogether. The bottom line is that corals will not grow in the seawater that will exist by the end of this century at the present rate of carbon dioxide build-up.

When writing *A Reef in Time* I had to struggle to describe ocean acidification, because the whole process is easily shown by simple equations but not so easily expressed in words. 'No equations,' my editor insisted. I don't think most of those who read my book would agree with that, but the book did carry the message, at least I hope so.

The year the book was published, and perhaps because of it, the Nature Conservancy held a workshop in Honolulu about the effects of ocean acidification on reefs. It was attended by Dick Feely, an expert on ocean carbonate chemistry whom I'd not met but knew well from his publications. I had many questions for him, but before I had a chance to ask any he said he wanted to talk to me about my book. I assumed I had gone wrong somewhere, but the following day Dick said that when he'd started reading he couldn't stop, and had read on all through the night. 'And the next day I started reading it all over again!' I felt flattered, but the point he was making was that, as a carbonate specialist, he hadn't realised the many ramifications of the subject. Ocean acidification is indeed seen in the geological record everywhere, and in coral physiology and in reef ecology. And I believe it will soon be a part of human history.

The cause of the mass extinction that finished off the dinosaurs is the most debated subject in the history of geology, yet curiously nobody recognised the role that ocean acidification might have played. As with so many sciences, all it needed was a little lateral thinking. Now acidification is taken much more seriously. It doesn't have the many complexities of climate change and so it receives much less coverage in the scientific literature. It is also less immediate as it won't be catastrophically affecting all the oceans for decades to come. Nevertheless, its consequences are irreversible on time scales far longer than those of climate change.

I fear that ocean acidification will be the ultimate evil of the Anthropocene – the only geological interval created by living organisms, us – and the primary cause of a mass extinction in the not too distant future.

One morning in June 2009 I had a phone call from a man who introduced himself as Paul Pearce-Kelly and said he was from the Zoological Society of London. He wanted to talk to me about *A Reef in Time*.

'Sure, Paul, what do you want to know?'

'I mean I want to come to Townsville to see you about it.'

'Oh. When?'

'Next week.'
'From London?'
'Yes.'

Paul certainly kept a tight schedule. The day after he arrived at Rivendell he was off somewhere else, having left me surprised by one of his questions. 'Charlie, what would you say if I arranged for an emergency meeting at the Royal Society, had Sir David Attenborough chair it, and had your talk streamed around the world?'

This guy's a nutter.

'That'd be nice, Paul.'

A few weeks later he phoned me again. 'It's all arranged for July the sixth,' he said. 'David's happy to chair it. We'll have a conference beforehand. He'll chair that too. What do you want to call your talk?'

'Really? Um, how about "Is the Great Barrier Reef on Death Row?"' I was thinking of Mary's younger brother Clive, a lawyer who'd had spectacular success stopping Americans from executing their own.

'Good,' said Paul. 'We've got a lot to talk about. I've booked you and Mary in at the zoo. They've got a nice flat there. Come as soon as you can. It's interesting to be with the animals when everybody else has gone home.'

On arriving at the zoo a few weeks later, Mary and I discovered that Paul, far from being nuts, was a ball of energy, self-effacing to a fault, and totally committed to conservation. The Great Barrier Reef Marine Park Authority had helped build a rather slick PowerPoint presentation for me. All that was needed was for me to write some introductory notes about myself, as requested by the Royal Society, so that Sir David could introduce me. I did as asked, omitting mention of my first visit there, after the Stoddart expedition! But I needn't have bothered – Sir David gave me his own introduction, and a generous personal one it was.

On the morning of my talk there was a large meeting of climate change and reef scientists at the Royal Society, with much of the debate being about what level of atmospheric carbon dioxide would give corals optimal environmental conditions. These views were subsequently published in a much-cited article, but at the time I said as little as possible as I had a threatening throat infection and was in imminent danger of losing my voice.[42] Fortunately that didn't happen: my presentation lasted an hour and a half, as arranged, and was followed by an hour and a half of questions, which only ended because the master of ceremonies called it quits. The conference, my talk and the questions had taken six hours; I was astonished at Sir David's stamina, as by then he was over eighty.

Some scientists thought my predictions about future carbon dioxide levels were overly pessimistic: they were at the extreme end of the Intergovernmental Panel on Climate Change predictions. Now, nearly a decade on, one slide from my talk gets aired by the occasional investigative journalist. It was made from data I borrowed from oil industry reports and publications, and it said that by 2015 we would have 400 parts per million of atmospheric carbon dioxide (we did). I extrapolated that this would cause major weather events (which happened), and severe bleaching, mainly during El Niño cycles (which also happened). This makes a clean rebuff for those denialists who claim that scientists make it all up, or (as I was sometimes accused) exaggerate. There is much to be said for such reality checks.

Naturally, I wondered how much difference my book, this symposium, and the spreading of what I had to say would make. For a long time, it seemed not much, but I later had cause to be more positive. As months turned into years I began to appear in more and more documentaries and interviews. I had made the effort to understand the many sciences involved with climate change and have been able to put that understanding to good effect.

Of all the articles, seminars, interviews and documentaries I have written, spoken and appeared in, one stands out in my memory. In 2009 Robyn Williams, who thirty years earlier at AIMS had recorded me talking about the first Australians living under coral reefs, asked me to do an item for *Occam's Razor*, another of his science productions for Australia's national radio. The production required a written script and had to be recorded in a studio. This was something new for me and I discovered how important speaking to a person is. I was in Townsville with only a microphone for company and the producer was in Sydney. He told me to take my time, and then proceeded to stop me about ten times in as many minutes.

'How did it go?' asked Mary, who had helped write the script.

'I haven't the vaguest idea,' I replied in all honesty.

When Robyn introduced my talk a couple of weeks later, he said it was among the most important he'd ever recorded, which is saying something coming from the person who probably knows the pulse of Australian science better than anybody.[43] It also says something about how times have changed, for what I said then would be old news today.

It was a plea from the heart, but trying to keep emotion out of it I had painted the future of coral reefs just as science said it would be, without spin or exaggeration. It's an incredibly important subject, yet now I'm reluctant to talk about it at all, for it's hard to see a bright side, and hope is an essential ingredient for just about everybody. Better to shut up and try to figure out why religions succeed when reason doesn't.

Our French hideaway had turned out to be one of Mary's great ideas for our family, but the outcome – my understanding of climate change – has robbed me of much of the enjoyment I once had in working on reefs, for I have a clear idea of what lies ahead, and wish I hadn't. It's an exceedingly sad thought that nobody will ever study the world's coral reefs as I have because much of that world no longer exists.

A very big website

Not long after *Corals of the World* was published I started thinking that the world needed a website where anybody could find out about all corals from all points of view – could identify them, and see a lot of photos of them and maps of their distribution. Advances in computer technology made such a website possible, but doing something about it was another matter. There were two main jobs: getting the mass of taxonomic information and photos together, and doing the mapping.

The ideal way to build such an all-encompassing website would be to have all taxonomists agree about the taxonomy, then contribute all they know. Unfortunately that will never happen, for taxonomists are an independent lot and they tend to battle opposing views. Not overtly, but through synonymies. It's a territorial thing, and can have its amusing side. But on a practical level, Nature's fuzziness does not make consensus easy, particularly for work that has a restricted geographic scope. Moreover, often what we read is as much about the taxonomist's personal biases as the actual taxonomy.

During the expeditionary work on the Coral Triangle, I'd come to know and greatly value Emre Turak and Lyndon DeVantier, field workers who excelled at recognising species underwater and who made careful collections for subsequent studies of species of interest. I asked them if they'd join Mary and me in building a website, which would be like the book *Corals of the World* but more comprehensive. Mary came up with the notion that users could get the website to build customised maps, which could be continually updated by us. It was a great idea and it worked in ways we never imagined, producing statistics and allowing all manner of interesting analyses to be made.[44] At least, that was the happy ending; the road along the way was more a matter of dogged perseverance.

In 2008, about a year after the Coral Triangle Initiative was launched, I was asked to attend a supposedly high-level conference about it sponsored by the Australian government, which up to that time had contributed little to its establishment. Naturally, given Australia's previous lack of interest, I wondered what the conference was all about, and am still wondering. I wasn't alone there: a group of delegates from Indonesia asked me to bring a coral triangle to the conference, as they'd never seen one. I explained that I only had one and it was too big to move. They said they never realised that . . .

There was also a mob from AIMS at the conference and they too seemed none the wiser, for they wanted to know why I was repeatedly mentioned whenever someone made a speech. 'If this work was your doing, Charlie, why wasn't it an AIMS project?' asked one.

'Well, someone suggested that but it got turned down. Remember?'

'Vaguely. So why don't we make it an AIMS project now?'

Why would I want to do that?

'If you guys did the website building, *maybe* we could,' I said.

Two years later AIMS's lawyer finally put the finishing touches to a thick document which said that the building of Corals of the World Online, as they called it, would be a joint operation: AIMS would do the website engineering, Mary would design the whole thing, and Mary, Emre, Lyndon and I would do the coral work. AIMS and I would both seek funding for it.

All went well until the AIMS scientist responsible for managing the institute's side of the bargain found another job; then their guy responsible for the mapping left also. We muddled on for a while, with computer technicians putting together a website with temporary text and photos from my book as placeholders. But AIMS was happy with this, corals not being high on their priorities, so we parted company.

Building the website we wanted would be a costly business; I screamed for help. Fortunately, Dave Hannan, the founder of the conservation foundation Ocean Ark Alliance and a character generous beyond belief, launched a financial rescue and Google Earth chipped in. By the end of 2011 we were up and running again. I also found a company who could engineer it all, with genuine skill and dedication.

The project just kept getting bigger, better and more complicated, and was soon costing much more than I'd anticipated, partly because it had no precedent and so didn't fit any mould. That also made it look like a risky venture, but nevertheless more generous backers who believed in what we were doing heard my call, and hundreds of others helped in all ways possible. We felt very privileged; we still do.

Despite the complexity of the undertaking, anybody looking at the website now would be forgiven for thinking it's all straightforward. It was designed to be this way, and to give instant answers. We still have a long way to go with it, for there's nothing to stop it having the capability to track the fate of all species, endangered or not, thus giving the best scientific support for their protection. This is a distant goal but would be the biggest step forward in coral conservation ever. It would be another massive task on top of all we have so far done, requiring detailed information about coral habitats worldwide and the paths of ocean currents that connect them, along with mountains of information about the biology of each species.

For me, an all-abiding problem remains. The website gives each species a name and includes descriptions, photographs and a distribution map. Species are therefore treated as if they're isolated units, even though, as I've explained, this is seldom the case. There are no options here, or at least none that I can think of, and it means that every point of information we have may not be exactly correct for a particular place. This may involve a minute error here, but a potentially larger one there. Such is the troublesome truth of reticulate evolution. I don't know if reticulate patterns are the last frontier, as some have said, but the subject is something that must eventually be confronted.

On 20 June 2016 a preliminary version of our website was launched at the International Coral Reef Symposium in Honolulu, amid 2000-odd delegates.[45] Having it ready was a cliffhanger: Mary flicked the switch at 4:30 am the day we left for Hawaii.

I pause to wonder what John Wells would think of our website. I hope he'd think his wall chart has been given a good future.

Names that matter

For more than two hundred years before the age of scuba diving, people had been amassing corals during expeditions of discovery to the tropical world; in fact, corals became one of the most popular collectables of all time. Naturalists collected them from reef flats at low tide, sailors who could swim collected them from deeper water, or they were purchased from natives with an eye for trade. By such means they accumulated in great quantities in museums across Europe and America, where they made contributions to natural history, especially when they were the subject of those big scholarly monographs that so defeated me when I first worked on corals.

Some corals in these collections were swapped or borrowed, sometimes returned and sometimes not. Inevitably, many specimens were lost, or given new labels when incorporated into another collection, commonly without any indication of their original source. No matter, except for type specimens, many of which were hard to track down if they weren't distinctively marked. Others presumed lost may never have existed. Of greater concern, type specimens were often oddities because they were atypical growth forms found only in shallow water.

Worse still, they did not represent any group or population – they were just different. This is ancient history but it's a history we must deal with, because it has created so many problems for nomenclature.

Coral taxonomists of the remote past, not being scuba divers, had no idea how species appeared underwater, and as we've seen, specimens were sometimes proclaimed a new species and given a new name simply because they looked different. Scuba diving opened a door to another world, and quickly became essential for reef field work, something Cyril Burdon-Jones foresaw when he applied for the grant that supported my post-doc. No doubt my scuba background was the reason I was offered a post-doc at all, given I had no formal qualifications in marine biology. No complaints on that score; I had the best training a would-be taxonomist could ever have – none. This enabled me to see central issues afresh, without being burdened by the baggage of the past.

A coral taxonomist working in a museum or university without studying corals on reefs would be in the same predicament as a tree taxonomist shut inside a herbarium without venturing into forests. What would Henry Bernard, who described so many species, have thought if he could have spent just one day on a reef? His world would have been turned upside down. Instead, having never laid eyes on a living coral, he abandoned binomial nomenclature and took up religion. Another author, the Indian George Matthai who published a massive volume in the same monograph series that much of Bernard's work was published in, ended up shooting himself. Maybe if those guys had spent a little time on a tropical island they might have lightened up a bit.

Seeing corals living on reefs allowed species to be identified with much greater certainty; it provided distinct criteria for separating closely related species that occur together, and revealed how variations in skeletal structures are linked to the environment in which the coral had grown. This reinvigorated coral taxonomy and led to detailed studies that, forty years on, are still progressing, providing a solid foundation for reef studies as well as overwhelming support for conservation. Nevertheless, we are still left with the legacy of the past, one which has more to do with the foibles of humans than with corals. This situation is far from unique to corals, but the ramifications seem particularly unfortunate in their case.

There are several types of coral taxonomists, the most prominent historically being the museum sort. Working on specimens collected by someone else, perhaps from unknown places and usually from unknown habitats, is not conducive to much original thought, so these people made the most of what they had by describing selected specimens in great detail and by constructing elaborate synonymies. In times long gone, such people tended to look to museum shelves for inspiration, but when all's said and done, Nature hardly got a look in.

Acropora (Acropora) hyacinthus (Dana, 1846)
Synonymy

Madrepora hyacinthus Dana, 1846; Brook, 1892.
?*Madrepora spicifera* Dana, 1846; Brook, 1892.
?*Madrepora surculosa* Dana, 1846.
Madrepora patella Studer, 1878; Brook (1893).
Madrepora conferta Quelch, 1886; Brook (1893).
Madrepora pectinata Brook, 1892; Brook (1893).
Madrepora recumbens Brook, 1892; Brook (1893).
Madrepora sinensis Brook, 1893.
Acropora pectinata (Brook); Vaughan (1918); Hoffmeister (1929); Thiel (1932); Crossland (1948); Nemenzo (1967).
Acropora spicifera (Dana); Vaughan (1918); Matthai (1923); Nemenzo (1967).
Acropora hyacinthus (Dana); Hoffmeister (1925) (*pars*); Thiel (1932); Wells (1954, 1955b); Stephenson & Wells (1955); Nemenzo (1967); Pillai & Scheer (1976); Wallace (1978).
Acropora conferta (Quelch); Wells (1954); Zou (1975).
Acropora corymbosa (Lamarck); Stephenson & Wells (1955).

A typical synonymy (this one from the author), where species and the taxonomists who use them are listed to indicate that the species are all one and the same.

The field worker is a very different sort of taxonomist. Spending a great deal of time underwater rather than in museums or laboratories is a much more difficult undertaking, and not just from a practical point of view. On the one hand, field workers must depend on the taxonomy of people of yesteryear who had very different experiences, then blend that into what they see alive on reefs. On the other, they constantly see corals that are new or different from any that have been recorded. These are the people I like to work with, for they usually have original knowledge and of necessity know a great deal about corals.

Molecular taxonomists, who work with DNA, are another breed altogether and today outnumber all others many times over. The best of them combine significant field experience with molecular technology and knowledge of the morphological characters of the species they study, and they seldom fail to get interesting results. But in this approach, good work is sometimes mixed with bad, because many researchers who've mastered molecular techniques have little field experience. A lot of students are in this business, striving to get to the front of the pack, and it's tough, with few winners and many losers. But there's a big future here, when molecular results are well integrated with a thorough understanding of the species in the field, their life histories, physiology and taxonomic background.

Many scientists try to be part of a combination of these different approaches, and now we commonly see co-authored publications where one author may have no idea what their co-authors actually do. Such works often read like something put together by a committee; however, the combination may tick the boxes of the editors of science journals, and that, in the pressures of today's academia, is usually the aim of the game.

Does it matter that the name changes these different approaches generate go ever on, with no end in sight? Yes it does, because it's the name that links information together, whether that information is in the form of a map, photo, description, survey, or any aspect of science or conservation involving species. Of recent years, taxonomy has been a popular subject for PhD theses, a thin (apparently deficient) thesis being one that has re-examined all relevant problems and found little cause for change, and a thick (apparently good) thesis being one that changes everything possible. The same applies to many publications, where the main aim is often change for the sake of change. If it is the end user that matters, names need to be stable and dependable, and, most important of all, the species that the name represents needs to be recognisable in the field and distinguishable from all other species.

That said, the real crunch has yet to come: reticulate evolution says that species divide and merge over big geographic distances as well as over evolutionary time, yet most computer programs, as I've pointed out, do not allow lineages to merge, they can only make branches. That's not something to ignore. Other problems are waiting in the wings, the most important being that some molecular results are clearly at odds with every other line of evidence relating to a species' phylogeny, and have led to highly unlikely speculations on the part of their authors. Such results are now being found in most groups of animals and I've not yet heard a general explanation for them, except for a possible role for ancestral junk DNA. So now coral taxonomy is not just about corals and taxonomists, it's also about molecular and information technology.

The whole subject has become so very multi-faceted, yet is still beholden to those collectors of ancient times and the specimens they deposited in museums. Type specimens and the name games they inspire were bad enough for field-going scientists like me, but are much worse for geneticists, whose primary expertise is in molecular technology and not the history of museum specimens. The DNA that molecular taxonomists use comes from living tissue, something not found in museum specimens.

Clearly this issue must be addressed: the old nomenclatorial system needs a drastic overhaul or it will become irrelevant. In the meantime, it's likely that chaos will prevail, which is fine for students keen to publish, but not so fine for anybody requiring names they can rely on.

Taxonomy in the minds of most people conjures up images of old museums, long boring monographs and even more boring people. But what is taxonomy? It is a description of how Nature is organised, one of the deepest and most complex subjects of all biology. Hardly boring, at least to my mind, but then I don't often think of myself as a taxonomist.

Pat Mather, who, you might recall, was aghast at my becoming involved in taxonomy, gradually changed her mind and eventually became my staunch supporter. But sometimes I think she might have been right in the first place, given the problems I've come across with the International Commission on Zoological Nomenclature.

Since its formation in 1895 the ICZN has produced, among other publications, the *International Code of Zoological Nomenclature*, which proclaims what taxonomists should, shouldn't, can and can't do – in some areas but not in others. This code gets revised every couple of decades or so, but it seems to me to be out of touch with taxonomy. Issues with the ICZN are critically important for natural scientists, although some don't realise it, or don't want to know about it. Certainly, these problems are not adequately aired. Like the frog that doesn't notice the warming beaker, we become acclimatised; we even think there's something normal about it, until it's too late.

It's true that the ICZN has done much to tidy up the chaos it inherited more than a century ago, but that tidying was done at a time when the central axiom of all taxonomy was that name changes should only be made if they increased certainty. That notion still exists in theory but has too often fallen by the wayside in practice, leaving us with many problems that should have been resolved long ago. These problems belong to every taxonomist, not just the ICZN, and all biologists who rely on species names would do well to be aware of them. Here are the most conspicuous:

Problem one: when a species name is changed (the old 'priority versus stability' argument), much of the information that goes with the old name gradually fades away. And as I've said, linking information about a species is what names are good for. Worse, the older the name, the less clarity it usually has. In many cases the oldest name is not even based on a type specimen, but on only a drawing or vague description, as with *Pocillopora damicornis*. Even today, taxonomists will triumphantly change an old, well-known name for an even older name that has never been used, and there's nothing to stop them doing this.

The name *Pocillopora damicornis*, for one of the world's most common and most studied coral species, is based on this rather doubtful drawing.

Problem two: one size fits all. Corals are not sponges, salamanders or parrots. These animals have almost nothing in common except the rules by which they are named. Corals have a special need here because so much work on them is becoming DNA-based. We need new type specimens with living tissue preserved: an essential case of 'out with the old and in with the new'. Certainly there are difficulties in doing this, but that doesn't mean they can be left in the too-hard basket.

Problem three: Latin! Latin was once firmly entrenched in the language of international law, religion, history, astronomy, anatomy, taxonomy, and heaven knows what else. Now, as far as I can see, it's only entrenched in the Roman Catholic Church and the ICZN. My book *Corals of the World* (not to mention many of my other publications and those of other authors) has descriptions of a large number of new species. All, says the ICZN, must be in correct Latin, whatever that is, for I don't know Latin grammar but am reliably told it has several forms. But I do know an old Catholic priest, unlikely though that seems; however, he confessed he didn't know any Latin either. So, heretic though I am, I did try, but for what reason? After all, we're only dealing with *two words*.

As things now stand, ICZN rules require a species name to be changed if that species is reassigned to another genus that has a different gender: how's that for creating a snag in today's world of electronic information searches?

Problem four: ancient junk. A long time ago, a German naturalist by the name of Lorenz Oken (1779–1851) devised an elaborate though mindlessly pointless system of nomenclature that included names of seven common coral genera we use today. The problem was that he didn't believe in binomial nomenclature, so the ICZN ruled them invalid ('unavailable', in their lingo). Six have since been approved, but poor old *Turbinaria* remains out in the cold. Who, besides me, knew that? Who, including me, cares?

This matters because coral taxonomy is riddled with such cases, the fate of *Favia*, one of the world's most commonly used coral names, being the latest.

When I first visited John Wells, back in 1975, I complained that the type species (the species a genus is based on) of *Favia* was obviously not a *Favia* at all. He agreed and commented that Thomas Vaughan (1870–1952), John's autocratic mentor, thought so too. Let sleeping dogs lie, John advised. And so I did, that is until *Favia* was given another name, one nobody had ever heard of. Why? Because *Favia*, one of Oken's once invalid genera, did not have a designated type species. A.E. Verrill, always on the lookout for something to do, gave it one, in another instance of Verrilliana – he got it wrong. So one may well ask: should an obscure 200-year-old publication, supposedly corrected by a hundred-year-old mistake, matter when the name *Favia* has now been used unambiguously in more than a thousand publications? Obviously not. I am not the complaining sort, but this time I couldn't resist publishing a complaint, using the politest phrases I could muster.[46]

I have been grumbling about the role of technical trivia for a very long time, my point being that the ICZN has done little or nothing about it. The account of *Favia* is an example but of these there are many, all contributing excuses for making changes ahead of reason.

Problem five: The multiple needs of molecular taxonomy. Enough already said, especially about the usefulness of old type specimens that don't have any DNA.

ICZN technical trivia reached a low point for me a year or so after *Corals of the World* was published, when a curator at the Smithsonian wrote to the journal *Science* claiming that all the new species I'd included in my book were invalid because I'd only nominated type specimens in a subsequent monograph, rather than in the book itself. As it happened, that same year (2000) the ICZN published a revision of their *Code*, which would indeed have left my new species invalid. The issue according to them was: which came first, my book or their code? After months of deliberation, which was actually between the commission and a friend of mine on the commission who took my side, it was finally agreed that my book did, but only because the contract I had with the printer was signed in December 1999, whereas the new code came into effect in January 2000. That is to say, had the contract to print my book (a matter involving a lot of time and money) been delayed a month, the commission may well have declared all the new species in it invalid. In the end the commission couldn't help writing to me, noting with approval 'my' use of their language (ha!), but also giving me a lesson in Latin grammar. What a lot of bunkum.

Decades ago the ICZN had essentially done the job it was originally created to do. If it can now shape up to future needs, excellent, for that will mean giving more attention to forward-thinking young people and less to worn-out history. Perhaps the tiger should give itself a new set of teeth.

The ICZN's rules pale into insignificance compared to the problems created by the Convention on International Trade in Endangered Species. CITES was established in 1973, and a very good move it was. All was well until 2002, when CITES's Animals Committee produced a list of coral genera that they considered unidentifiable. They further decided that it was bad to collect corals for the aquarium trade: what if endangered species were being collected? As corals were too hard (for them) to identify, they simply classified all scleractinian corals as endangered species. The precious deep-water coral *Corallium*, which is not scleractinian but *is* threatened, having been harvested without control for centuries, did not make the list – that was blocked by commercial interests. The aquarium trade, which does immeasurably more good than harm, is battling on with uncertain success, but not so the scientists who need to collect corals for their studies, and to do so in all countries where the species they are interested in occurs.

The people who manage CITES in Australia and the US found ways to bolster their regulations, the most effective being that an export permit from the country of origin must be obtained before an import permit can be issued. That seems straightforward, but when I first applied for an export permit some countries that have corals had never heard of CITES. Still, customs officials learned how to exploit that quickly enough.

'You want export permit? Export permit take lot of work. Cost extra for quick processing.'

'How much extra?'

'Forty American dollar. One hundred dollar better. Cash only.'

At first Australian Customs were not too concerned about corals. I would just explain who I was and what the corals were for. Then they tightened up. Now they will pounce on anything that looks like a coral.

Corals are not made of ivory and corals are not white rhinos. Certainly they are endangered, due to mass bleaching caused by climate change, but the CITES charade has done enormous damage to coral research everywhere while doing nothing meaningful for conservation.

In Retrospect

The underworld

It's hard to believe, but when I started diving I knew nobody who'd ever been scuba diving, except those French photographers I met on Heron Island. Now there are at least six million recreational divers worldwide and they spend most of their time on coral reefs. But diving for research is not like diving for any other reason: scientific divers usually go to places seldom visited by others, remote places where the water may be murky, the currents unpredictable and local knowledge non-existent. Nor are these divers there to enjoy the scenery or take snapshots, they are there to work. To do this safely requires a lot of know-how and careful thinking, especially when handling equipment underwater. The scientific diver must concentrate on the job, and so the actual act of diving needs to be second nature. And this kind of diver must know how to deal with the unexpected and must not depend on others to make decisions.

My style of diving, where I decide what's safe and what isn't, has long ceased to be compatible with how things are done at most research institutions the world over. I have no problem with regulations if they're based on knowledge and reason, but sometimes it seems that a learning curve is missing from institutionalised diving. Some of the regulations created by bureaucrats in my time are fascinating, such as keeping unbroken, eye-to-eye contact with your buddy. Even blinking would have violated that one, and doing any work would have been impossible. Another brilliant idea was that a fully kitted up rescue diver be on standby during all dives. In the height of summer on the Great Barrier Reef it would have been the rescue diver who needed rescuing, from heat exhaustion. Then there was the notion that divers had to have a surface buoy attached to them by a float rope – which would get entangled with every turn around a coral or propeller on an outboard motor.

These rules and regulations kept changing like the proverbial tide, and many were seriously daft. Since that first nervous dive in Sydney Harbour fifty years ago, I have logged more than six thousand hours on scuba, most of them on the world's great coral reefs. This might sound like six thousand hours of fun, but almost all my diving has been for work that requires heavy-duty thinking. And that, dare I say it, means diving alone. This is not something peculiar to me; just about every diver I've known whose job involves a lot of concentration feels the same. We might have a helper along, in my case to carry a basket for the corals I collect, but if so I'm not looking after the helper, which means they are diving alone.

Amateurs are usually aghast at the thought of not having a buddy when diving; it breaks rule number one, they say. Sometimes on a dive from a regulation-infested boat, rule number one just has to be obeyed, so it's good to team up with someone who also has real work to do, then the rule becomes 'same time in the same ocean is a buddy dive'. All we need do is start the dive together, then go our separate ways.

Unfortunately, this hasn't always been so easy, especially those times in Japan when I was hosted by the owners of dive shops: then my buddy was invariably a son or friend of the shop owner.

IN RETROSPECT

As most reefs in Japan lie in the path of the Kuroshio, working on them usually involved swimming down to the bottom quickly and then crawling along, hand over hand into the strong current, holding onto chunks of reef or coral. When so doing, I often found myself with a buddy swimming frantically above me in the full force of the current and rapidly getting exhausted. On four separate trips my buddy had to be rescued and it became obvious that it was only a matter of time before a rescue would be a serious, perhaps fatal, matter. Had it been possible to talk about the dive before we did it, we might have been able to sort it all out, but most of the time I found myself with a buddy who couldn't speak English, or if he could, would not admit that he wasn't up to diving with an old *gaijin*. Diving alone, as I always wanted to do, was much too dangerous, they insisted.

After one rescue that could indeed have been serious, because it turned out that my buddy was a novice, I hit upon the idea of trying to 'drown' new buddies on the surface before the dive. This was done by enticing him to go for an ordinary swim, out into the current. Most Japanese don't swim well, or didn't in those days, so when a would-be buddy had to turn back I had a good excuse to go it alone. To save face – his face – I would give a shamefully exaggerated account of my swimming prowess, after which I was generally left to get on with it alone.

For me, rule number one is usually 'for safety's sake, dive alone'.

Dive computers have been the greatest invention of all time for scuba divers, yet incredibly, some organisations still insist on their divers using only tables to determine how long they can remain at a given depth. Of course computers, like any instrument, can fail, so experienced divers always keep a mental tally of how deep they have been and for how long. That's good practice, but fantasy tables that effectively stop a diver from doing much more than paddle about in the shallows are pointless. Before dive computers were invented I used US Navy tables, but for me even these were over-conservative, for I knew I could at least double the bottom-time they allowed. That's not to say my times would be good for everyone, but they allowed my dives to be many times as long as table-loving institutions now allow. If that's recklessness on my part, how come I've never had any trouble in that regard in so many dives?

I have seldom felt that my regular work diving is hazardous, except perhaps when decompressing. This is normally done by holding onto a rope dangling from a boat for perhaps an hour or more and is terribly boring. Twice I have fallen asleep and discovered that there's no awakening quite like breathing in seawater. But I'll admit that this sort of diving can sometimes be dangerous if done in remote places and using unfamiliar equipment in unfamiliar waters without local guidance. So be it; I understand the risk and accept it. I do insist on having a watchful boatman on hand if there's a current; that's something I've come to be fussy about.

Perhaps the most memorable diving I've ever done has been on the outer reef slope of a ribbon reef at night, with a full moon. This needs careful planning, and to be kept secret lest somebody else wants to join in. The first step, done surreptitiously during the day, is to anchor a little buoy on a good spot well away from the mother ship and tie a cyalume lightstick to it. Then, when everybody has gone to bed, you sneak away with a zodiac. The rest is simple: find the buoy, anchor, and slip over the side. With strong moonlight the reef is easily seen, although it's best to use a torch to find a really good spot. That done, the dive is just a matter of turning off the torch, lying back, and relaxing.

In very clear water the whole reef is a spectacularly beautiful silver-grey and full of life. It is noisy, busy and utterly surreal. Looking up, you see schools of fish flashing silver in the moonlight on one side, then deep grey on the other as they circle. On the bottom, crabs and lobsters crawl ghostlike among the sea cucumbers, urchins and starfish. Corals, seen mostly in silhouette, have outstretched tentacles that make all manner of otherworldly shapes. Some fish are asleep, tucked in the coral and enveloped in a mucous cocoon. Others are hunting. Sharks appear out of nowhere, perhaps to circle around to check out the intruder, then vanish into their silver world with a burst of speed. They too are hunting, giving further spice to an already overloaded atmosphere. Before it's time to return to the zodiac it's good to spend a little time swimming around with the torch turned on, because the beam floods everything ahead with colour while the rest of the world goes dark. Reefs seen by moonlight when you're alone are thrilling places; it's an experience never to be forgotten.

Some divers might think that such junkets are just reckless, but I don't. Diving is dangerous if you believe you'll be rescued should something go wrong. Good preparation is everything.

In September 2014, the American Academy of Underwater Sciences bestowed their Lifetime Achievement Award on me, making me the first non-American to be so honoured. At the award ceremony, I ended my talk with a description of diving alone on the Great Barrier Reef in the moonlight. With dozens of top-ranking diving officers in attendance I thought it might not go down too well, but disappointingly they gave me a standing ovation. The ceremony was held at Sitka, Alaska, a place way outside my diving experience, and I'd signed up for a post-symposium, drysuit diving tour, as I wanted to see the giant kelp beds I'd heard so much about. My participation was assured provided I submitted a pile of paperwork the likes of which I'd never imagined, including a letter of approval from my diving supervisor. Diving supervisor? I think I was the only person at the entire symposium who wasn't allowed to go diving.

'Hey man, those moonlight dives sound real cool,' said one diving safety officer after my speech. 'That's just what I'm gonna do if I ever get to that great big reef of yours.'

'You'd do that? What about all those bloody awful regulations you guys have?' I said.

'Ah, that's just paperwork, man. No problem.'
No problem for you, old buddy.

A year or so later, after Mary and I had launched our website in Honolulu, I took the opportunity to go on a rebreather dive with a couple of friends. After an hour on a fast boat we found a suitable spot near the entrance to Pearl Harbor. Rich, the dive leader, gave me a good briefing as I'd never used a rebreather before. Unlike with scuba, there are no bubbles; breathed air is recycled through a scrubber, which takes out carbon dioxide so the air can be reused. This allows divers to go much deeper and stay down longer than is possible with scuba.

Down we went. All went well until, nearing the end of the dive, I felt something was amiss and looked up to check where our boat's anchor chain was. Then something hit me like a blast from hell. I inhaled a mouthful of chemical from the rebreather's carbon dioxide scrubber, which contains sodium hydroxide (caustic soda). I turned my regulator to scuba mode and then copped another mouthful, the same as the first. A lifetime of diving has made me immune to panic, but nevertheless the cover of Douglas Adam's *The Hitchhiker's Guide to the Galaxy* flashed before me: DON'T PANIC. In a haze of pain, I vaguely considered my options. Heading for the surface without air was out – lethal. Another breath? Just then I felt Rich's hand gripping my arm and, manna from heaven, the next breath from my regulator was oh so wonderfully, deliciously sweet clean air.

On nearing the surface my body protested with salvos of vomit and diarrhoea, but otherwise I thought my ordeal was over. I felt okay, but my optimism was short-lived as we headed full-speed for a hospital. With the boat pounding through the waves, my airways started swelling, and that kept worsening until I was barely getting enough air to stay alive. I maintained an icy calm, as I'd learned to do during the severest asthma attacks of my youth. On arrival at the hospital, still able to walk although hardly breathing at all, I clenched an oxygen mask to my face and lost all memory of the rest of that day. When I awoke the following morning I had a large metal breathing pipe down my throat and was coughing out of control. It hurt like hell, even though the hospital had, I learnt, bumped up my painkiller as high as they could without putting me into an induced coma. Mary told me I was in an intensive care ward. It appeared that half my epiglottis was missing.

The following day a specialist told me my throat had been badly burned and that I would probably have problems eating, breathing, and perhaps talking. Modern rebreathers had long been considered safe, but now it seemed that Rich's paradox, as he called it, had come true: the longer it seems safe, the more dangerous it actually is.

As the day unfolded there was no lack of ideas about how my troubles might be rectified. One was to insert a tube into my stomach and leave it there until I could eat normally. Another was to remove the breathing pipe from my throat and insert a plastic gadget directly into my windpipe so I could breathe through my neck. The latter was promptly done, bringing memories of Katie's babyhood flooding back in harsh reality, and it worked for me just as it had for her all those years ago. The plastic gadget was taken out shortly before I left the hospital a week later. By holding a finger over the hole in my neck, I could speak, albeit with a voice that made old Satchmo sound like a soprano. No matter, I made a complete recovery, voice and all, with only a little scar on my neck as a souvenir.

I'm glad my epiglottis was okay and I didn't need that stomach portal. Pouring a glass of red straight into my stomach while sitting on my swing chair at Rivendell would surely have ruined the bouquet.

Diving in tropical countries brings particular hazards, especially when working and living with indigenous people. This is an aspect of my life I've always treasured, but it also means continual exposure to a wide array of diseases, the majority of which are easily caught. I seem to have become resistant to most of them, which is fortunate because I'm also resistant to seeing doctors, but I've no resistance to malaria, a big problem in the central Indo-Pacific tropics. I've had it four times, each episode very different from every other. The first time, I had ordinary malaria and just put up with it. The second time it was the potentially fatal cerebral malaria, which I treated by climbing into a bath full of ice whenever my temperature went sky-high.

The third time was many years later. I'd been writing a book at home, and in the mornings I would feel ill and freezing cold, despite it being midsummer. I'd mutter about seeing a doctor but in the afternoons I'd feel fine and forget about it. This went on for over a week, until Mary pointed out that I could give whatever it was to our kids if I didn't do something about it.

My doctor agreed with me that it wasn't malaria, but she had no better idea what it might be.

If one must get a tropical bug, Townsville is an excellent place to do so because half the Australian army is based there, and all their doctors specialise in tropical diseases. 'It's malaria,' said the army.

I was bundled into hospital and three days later it was all cleared up, thanks to a new antibiotic. Then, just before I left the hospital, another doctor turned up and wanted a blood sample.

'No need,' I said cheerily, 'all cured.'

'I'm not here to cure you, mate,' he said, 'I want your blood for teaching – it's a bloody zoo.'

The fourth time I got malaria wasn't so bad for me, but it might have been for Australia. I had been feeling off-colour during a trip to America, and the flight back home was particularly tedious.

The plan was for me to join Mary and our kids in Cairns and head north into the rainforest. At the airport, we considered going back to Townsville, a few hours' drive away, but on we went. I hadn't visited a malarial region for well over a year, so it seemed extremely unlikely that that was the problem. By the time we reached Cooktown I was in bad shape, so we stopped by the hospital. It was a Sunday evening but two nurses came in to see me.

'Take some aspirin, love, and you'll be fine,' said one. The other looked dubious, so a blood sample it was.

I was a little surprised when the manager of the caravan park where we spent that night came around soon after dawn and said I had to go back to the hospital, urgently. The hospital put me in touch with the infectious diseases unit in Brisbane, who gave me the happy news that it was malaria again.

'Soak yourself with mosquito repellent and get back to Townsville as quickly as you can,' they commanded.

We duly went, after which I was on the state news, day after day, identified as 'the Townsville man'. Despite my being reasonably conscientious with repellent, mosquitoes had bitten me, hung around for the fortnight, then bitten thirteen other people, giving them my malaria. That's how the life cycle works.

'I'm feeling really bad about this,' I lamented after yet another long discussion with the infectious diseases unit about exactly where I had walked in the Daintree before reaching Cooktown.

'Well, don't be,' the guy said. 'It's twelve years since we had a case like yours – it's been fun.'

Fortunately, all those who got infected from me were from the south (belated apologies to you all), where there were no *Anopheles* mosquitoes, so there was no chance of it spreading further. But if the authorities hadn't acted so quickly, I could have started an epidemic in north Queensland.

There's no medical examination I've ever heard of that allows someone who's suffered serious asthma to go diving. Luckily for me, diving medicals hadn't been invented when I started out, and by the time they became fashionable – always the first step to becoming compulsory – puffers had also been invented. Several colleagues have had a similar history of asthma, and like me have had to fudge their diving medical. I'm an expert at doing that. I have talked about this issue with two medical specialists, and find their explanations – that the water and air mixture divers regularly inhale can trigger asthma – unbelievable. There's more than one trigger for asthma, with exercise and air pollution – the latter the cause in my case – being the most common. Air from a compressor that doesn't have a clean carbon filter gives me a headache, but not asthma. I've never had the slightest hint of asthma while at sea, and I wonder how many careers have ended for what, from the evidence I've seen, is likely to be a fallacy. Many experienced divers and the occasional instructor know exactly what I'm on about.

There's no shortage of animals on coral reefs that sting and bite, and I've been stung by most and bitten by many at one time or another. In the minds of most people sharks head the danger list, but I have only been seriously attacked by sharks once – in thousands of hours in their company. I love having sharks around: they're sleek, graceful, and exude the essence of primordial power. Yet, after a few obvious precautions are taken, they're far less of a hazard for scuba divers on reefs than dogs are for joggers on suburban streets. It's one of the great tragedies of our age that they have been so mindlessly slaughtered – just to make soup.

I'm struck by the good fortune I've had, diving in so many fabulous places and seeing so many things that others probably never will. I've helped photographers try to capture sights for others to enjoy, but how can anyone film a huge school of fish as they form a silver cylinder, slowly revolving, with one's self in the centre? After a few minutes the cylinder stops and it's you who's revolving, because there's nothing other than fish to see and the brain does a correction. There is no up, except the direction of air bubbles, and no down, except what the depth gauge says, until the bottom suddenly comes into view. That's gut-wrenching; for a moment the brain rebels, refusing to believe the bottom is fixed, then the fish resume revolving, and you're once again stationary.

Hardest of all to convey is the feeling of wilderness, of being completely alone in spectacularly beautiful places. I sometimes envy the early explorer-naturalists, with so much awaiting their discovery. But then they didn't have scuba.

Views from my coffin

Living at Rivendell with my family has been at the core of my existence. My 'little boy' now towers over me (and beats me at chess without trying). Eviie and Martin are both preparing for a future entrenched in the world of Nature, choosing career paths that are about giving, not taking. Rivendell has changed over their lifetimes, and over Mary's and Katie's, but it has kept the same ambience. Full of books and surrounded by trees and wildlife, it remains a place conducive to thinking – a perfect environment for the likes of me – but it also keeps prompting me to worry about my family's future, and that of our planet. I'm not unduly pessimistic but I cannot ignore reality, and that reality weighs heavily on me. I imagine that anybody who believed they could foresee the horrors of the last world war would have been in a similar predicament. I envy those who are free of such burdens.

At times of change – and clearly there have been plenty of them for me – I like to sit on my coffin, so to speak, and look back; it gives perspective. When doing this in times past I used to see myself as someone who'd meandered by chance from one career path to another, not someone who followed any sort of purposeful direction. Now I see the opposite. I was a naturalist as a child and have remained a naturalist ever since, earning my living by observing, reading and thinking, ultimately trying to make sense of it all in ever larger contexts.

I have propped my ladder against many walls but I've never climbed any of them. I have been immensely lucky. Other than my expression of interest in the post-doc at James Cook, I've never applied for a job, or a promotion. I've only ever had one master – the knowledge I sought, in whatever guise that came in – and I've never advanced my career if that advancement seriously compromised my independence.

I'm quite certain that my love of the natural world stems from the freedom I had as a child to be alone in places that interested me. I love the tropics, where I've lived for most of my life, and my affinity with coral reefs tops all. But my fascination with marine life started at Long Reef, at the age of six. Nobody taught me about that marine life. Everything I learned, even if from a book, was for me a personal discovery, stemming from curiosity and becoming hardwired somewhere in my head.

My connection to the sandstone country of Sydney's Ku-ring-gai Chase – the bushland I so enjoyed with Jinka – still goes deep after fifty years. I loved pondering its spiders, beetles, lizards, frogs, ancient banksias and ants' nests then, and I still do now. To this day I find it hard to walk past a rotten log or a patch of swamp without wanting to take a closer look. Maybe I'll find a strange flatworm, another sort of moth or gecko, or perhaps a weird fungus. Of course, any seashore draws me, just as it always has.

Looking back on my work with corals, one interval stands out in importance above all others, and that's the two years I spent as a post-doc at James Cook University. It was important because I had nobody telling me what to do, think or say, nor any pressure to get anywhere. I had the opportunity to travel and was free to think for myself. And while I couldn't rationalise then what I saw and thought about during those years, I can now. What I found was that the fundamental realities of Nature were to be discovered in the real world and not in books, laboratories, museums or, later, with computers. It can be too easy for modern technology to blind us to its failings if we don't spend time absorbing the realities of Nature for ourselves. Context is everything, and the real world must be experienced rather than just analysed with tools. Young scientists need to be encouraged to take the time to soak up Nature's realities, and to think. Only then can they meaningfully integrate technological approaches.

I always took that time, be it with dragonflies or corals. What I saw with corals was how they changed from one place on a reef to another, how species differed when they occurred together, and how all of this changed with geography. I realised that corals must have a great capacity for dispersal. We take such matters for granted now, but when I first turned to corals they were barely considered by anybody, and so just about all concepts of biogeography and aspects of evolution that it underpins were on untrustworthy grounds. To work on these fundamental questions I didn't need a research grant or to get anybody's permission or to make reports, I just needed to go diving and then think about what I'd seen, often as not for a decade or more.

Didn't all this make my job the best in the world? I was regularly told so. Certainly it was off the scale by modern standards, which makes me feel desperately sorry for today's students, who have little independence. They are plunged into a rat race that stifles real advances. Many still do well, but often don't reach their true potential.

There's no way the likes of me would thrive in today's funding-fixated, ResearchGate-scoring, occupational health and safety-obsessed, time-tracked, committee-controlled, paper-counting, regulation-saturated, supervised academia – the world of bean-counters. If I had been confronted with all that I would not have set about creating a global coral taxonomy and biogeography, would not have been able to produce big-picture syntheses, and would not have had the time to do the thinking that revealed fundamentals like reticulate evolution.

Most professionals today, be they in the sciences, arts, education, even sport, work within a cage of bureaucracy that controls most aspects of their working life. A few creative souls succeed but most don't. When I look at the résumé of a scientist, as I've needed to do every few weeks, I'm nearly always presented with a list of degrees, employment positions, awards, grants, collaborations, plenary talks, committee memberships, and so on – and on. Then I read about the field this person is in and what they're doing in it, and finally there's usually a long list of publications. All well and good and all part of today's ways and means, but often as not I'm left with an unanswered question: what has the person actually achieved? This prompts me to ask: how different would the world be if this person's work didn't exist? These questions would be easily answered by some, but uncomfortable for those whose PhD work was their only significant discovery.

The moments in my career I have valued most were when I discovered a new species or found out something previously unknown. About half of all corals were effectively unknown entities when I first came across them and most of these didn't even have a meaningful name. Sometimes I found an old name for them, sometimes I gave them a new one, and sometimes I set them aside for further study. This is not stamp collecting, it's creating new knowledge in the belief that it will be put to good use one day, by many people.

Today, most scientists fortunate enough to be paid for their work have the same problem that Lewis Carroll's Red Queen had: they must run flat out just to stay where they are, to service the many needs of the cage they're in. Academics grumble about their teaching load and the amount of supervision they must do; and research scientists complain about the long and complicated grant applications they must write (and others they must review), the papers they must 'get out' (and others they must review); and everybody resents the amount of time consumed by administrative chores, committees and meetings. Some are happy with this situation, for if the needs of their cage are met they can spend their whole careers in it, maybe getting promotion and tenure; that makes for a comfortable and honourable career even if little is actually achieved. For the motivated academic or scientist who needs time and headspace to think, something needs to change. It's not enough to point to sabbatical leave, that fleeting opportunity when they can escape from their cage to do what they should have been able to do in it. This is a system that selects for mediocrity, leaving others longing for the freedom they need to get on with what matters.

These sorts of issues have been around a long time, but it's only in the past decade or so that they've really come to dominate the careers of creative people, including those keen to do something to help rectify the damage humanity has inflicted on our planet. I resigned my job before the bureaucratic cage could trap me. Not everyone has had that luxury.

It was the impending demise of coral reefs that turned me into a media tart. I felt I had little choice in this and now, more than a decade on, I have no regrets. Science has done its part in identifying the path of destruction humanity is on; the job now is to get this message out to the wider world, by any means possible.

In November 2014 Sir David Attenborough and his retinue of ten filmmakers met for dinner at Rivendell, before the start of a documentary series about the Great Barrier Reef, screened worldwide in 2016. I had not seen Sir David, by then aged eighty-eight, since my talk at the Royal Society five years earlier. Once again I was much taken by his gentle, modest nature – my whole family was – and I loved hearing some of his recollections about people I remembered from the earliest days of television. Even so I was astonished to find out that he had filmed the life of Raine Island six years before I was there in 1973. So much for my belief that we had rediscovered the place. A week after that dinner I joined the team on Heron Island, where Sir David interviewed me on all manner of subjects. It helps to be relaxed at such times and I was; his interest in what I had to say and his disarming self-effacement made me so.

During 2016, with most of the world's reefs undergoing massive damage from coral bleaching, I was interviewed sixty-one times, chiefly about the Great Barrier Reef and probably to little or no effect in most cases. There were some exceptions, though. A talented film producer invited me to participate in a documentary about mass bleaching, primarily aimed at young people. The film, *Chasing Coral*, was launched at the Sundance Film Festival in January 2017 to resounding acclaim and was awarded the viewer's choice prize. Further awards soon followed. *Chasing Coral* has the potential to change the hearts and minds of millions of viewers.

For me one scene is particularly poignant. It was filmed on a floating restaurant in New Caledonia where revellers were living it up, drinking and dancing, yet just under their boat the coral reef was dying. Has it come to this? Is humanity really going to party on regardless of what's happening to the planet? I think not; the wall of ignorance that many have battered their heads against is crumbling. Even in Australia, the cries of climate change denialists are waning and a new era of thinking and caring is tangibly dawning. It is a change being led by technology and younger generations.

The older I get, the more people I meet who tell me that their love of the natural world stems from their childhood. When these people were children they were scientists, for they were keen to observe, explore and understand. Some stayed on that path and are now teachers or professionals, others went in different directions but remained amateur naturalists. All have a gift they were born with, the ability to appreciate Nature. What greater gift can a person have?

Despite the parallels between my life and that of my namesake, I flatter myself that I am most akin to Aboriginal people, in thought at least. They have survived this long because they have understood their land, have listened to it, been a part of it. Certainly they did not seek dominion over it. Like it or not, all humanity is going to have to learn that lesson, for the age of exploitation is all but over. We will either learn to protect and be part of what we still have or become some sort of alien life form on our own planet. Or go extinct.

As the saying goes, we seldom see things as they are, we see them as we are. While the view from my coffin has changed many times, one perspective has remained constant, and that is that every hour of my life I spent in a meeting was a wasted one, and every hour I spent with Nature has been blessed.

Afterword

Night comes quickly in the tropics. From my swingchair on the terrace at Rivendell, dogs at my feet, a book and a glass of wine at my elbow, I love to watch the last flocks of cockatoos, ibis, cormorants and lorikeets flying high overhead. Higher still, the pelicans glide effortlessly in V-shape formations. Then, almost out of sight, I see flocks of kites spiralling on steady wings as they catch a ride on the last thermals of the day. I like to listen to the chatter of the little birds as they prepare for sleep and to the frogs and crickets as they commence their nightly croaks and chirpings. To my right a couple of possums argue in their harsh guttural rasp. They are high in a fig tree that now overgrows the big rambling house behind me but which started life as a pot plant in a corner of my office at AIMS. To my left lies a long expanse of a small river, enclosed by trees and full of life. My view constantly changes as night comes, with the season, and with the light of the moon.

It wasn't always like this. Two hundred years ago I wouldn't have had to clear the riverbank in front of me of lantana and chinee apple, and I might have been wary of the locals, for they would have found me a very curious person, perhaps another invader. For thousands of years before that, little else would have

AFTERWORD

been different, but going back twenty thousand years the river would have been dry, the trees absent, and the locals would have been three or four days' walk to the east of Rivendell – where the coast was before the sea rose. Perhaps these first Australians might have ventured as far inland as my home on hunting trips, but mostly they would have lived near the outer edge of half a million square kilometres of savannah woodland, the place we now call the Great Barrier Reef. The young men would have hunted goannas, kangaroos, emus, and larger animals that no longer exist. Women and children probably remained on the coast, climbing the rugged escarpments that separated their limestone caves from narrow beaches below, perhaps to check on fish traps or to gather turtle eggs and shellfish. The air would have been full of the sound of waves, the squawking of seagulls and the laughter of children.

As darkness came, the day's bounty would have been cooking and the firelight might have revealed paintings and ochre-filled etchings on the cave walls. Maybe the cold night air would have been filled by the ethereal lament of didgeridoos and the chanting of songs telling of how the sea had once risen and destroyed their home but given another in its place. At that distant time, giant marsupials might have ambled down the bed of my river, keeping a wary eye out for monster birds of prey and, for a further forty thousand years before then, watching out for humans, the only other creatures they need fear.

Nowadays, night takes away the river if the moon fails and brings with it the dark thoughts that plague me. Super-cyclones the likes of which I have not seen before impinge upon us, and the river I love could come threateningly close, yet I know that future climates could see it run dry for years. Science has revealed the facts of the matter and they are starkly clear: there will be no better future without drastic action on climate change. I will not be taken by surprise, but most of my countrymen will. The worst heat, fires, droughts and floods on record have already taken place in my fig tree's lifetime, and in just that blink of an eye we have lost half the world's coral colonies, and reefs everywhere are stressed.

    I write these last words having just returned from a flight over The Reef and some diving near Cairns with a television crew. The reefs I saw, from both above and beneath the water, were all severely bleached. Now I fear that this could have happened to the entire Great Barrier Reef. I had imagined I would be gone before the worst of my fears were put to the test, but now I'm afraid this may not be so. Perhaps one day I'll be able to take solace in the thought that I've done my best to protect the place I so deeply love and which has played such a big part in a life more fulfilling than I ever imagined. But I suspect not.

Notes

1. Veron, J.E.N. (1973), 'Physiological control of the chromatophores of *Austrolestes annulosus*', *Journal of Insect Physiology* 19: 1689–1703 and (1974); and 'Physiological colour changes on Odonata eyes. A comparison between eye and epidermal chromatophore pigment migrations', *Journal of Insect Physiology* 20: 1491–1505.
2. Belatedly, after learning a bit about corals, Veron, J.E.N., How, R.A., Done, T.J., Zell, L.D., Dodkin, J. and O'Farrell, A.F. (1974), 'Corals of the Solitary Islands, Central New South Wales', *Australian Journal of Marine and Freshwater Research* 25: 193–208.
3. Wright, J. (1977), *The Coral Battleground*, Thomas Nelson, second edition (1996), Angus & Robertson. Judith, a writer and poet of great acclaim, was the daughter of P.A. Wright, whose family mansion Booloominbah I saw on my first day at the University of New England.
4. *ibid.*
5. The Great Barrier Reef Committee, founded in 1922, is the world's oldest coral reef society. Its main task was to establish and run the Heron Island Research Station. In 1988 its name was changed to the Australian Coral Reef Society, which has flourished ever since.
6. The Yabulu nickel refinery then dumped its waste into dams, turning them into one of the worst environmental problems in Queensland's history.
7. Purdy, E.G. (1974), 'Reef configurations: cause and effect', in Laporte, L.F. (ed.), *Society of Economic Palaeontologists and Mineralogists, Special Publication* 18: 9–76.
8. The southern Great Barrier Reef tells a different story, for there the Coral Sea is relatively shallow and is littered with fossil reefs now drowned by movements of the sea floor and the effects of eons of sea level changes.
9. Veron, J.E.N. (1978), 'Deltaic and dissected reefs of the Northern Region', *Philosophical Transactions of the Royal Society of London B Biological Sciences* 284: 23–27.

NOTES

10 Veron, J.E.N. (1978), 'Evolution of the far northern barrier reefs', *Philosophical Transactions of the Royal Society of London B Biological Sciences* 284:123–127.

11 Yonge, C.M. (1930), *A Year on the Great Barrier Reef: The Story of Corals and Their Greatest Creations*, Putnam, London.

12 Grosse, P.H. (1860), *Actinologia Britannica. A history of the British Sea-anemones and Corals*, Van Voorst, Paternoster Row, London, England: 362pp.

13 Veron, J.E.N. (2003), *Inge*, Glass House Books: 227pp.

14 Veron J.E.N. (1992), 'Environmental control of Holocene changes to the world's most northern hermatypic coral outcrop', *Pacific Science* 46: 405–425.

15 Veron, J.E.N. (1992), *Hermatypic Corals of Japan*, Australian Institute of Marine Science Monograph Series, 9: 244pp.

16 Nishihira, M. and Veron, J.E.N. (1995), *Corals of Japan*, Tohoku University (in Japanese), 439pp.

17 Wells, J.W. (1955), 'A survey of the distribution of reef coral genera in the Great Barrier Reef Region', *Report of the Great Barrier Reef Committee* 6: 1–9.

18 Veron, J.E.N. and Minchin, P. (1992), 'Correlations between sea surface temperature, circulation patterns and the distribution of hermatypic corals of Japan', *Continental Shelf Research* 12: 835–857.

19 Veron, J.E.N. (1986), *Corals of Australia and the Indo-Pacific*, Angus & Robertson, Sydney. Reprinted by Hawaii University Press, Honolulu, 642pp.

20 Veron, J.E.N. and Hodgson, G. (1989), 'Annotated checklist of the hermatypic corals of the Philippines', *Pacific Science* 43: 234–287.

21 Veron, J.E.N. (1993), *A Biogeographic Database of Hermatypic Corals: Species of the Central Indo-Pacific, Genera of the World*, Australian Institute of Marine Science Monograph Series 10, 433pp.

22 Veron, J.E.N. (1995), *Corals in Space and Time: The Biogeography and Evolution of the Scleractinia*, Cornell University Press, New York. 321pp.

23 Wells, J.W. (1956), 'Scleractinia' in Moore, R.C. (ed.), *Part F Coelenterata* (in the 23-voume series *Treatise on Invertebrate Paleontology*): 328–444.

24 Veron, J.E.N. and Kelley, R. (1988), 'Species stability in hermatypic corals of Papua New Guinea and the Indo-Pacific', *Association of Australasian Palaeontologists Memoir* 6: 69pp.

25 The condemnation was in Heck, M.K and McCoy, E.D. (1978), 'Long-distance dispersal and the reef building corals of the eastern Pacific', *Marine Biology* 48: 348–356. The proposal was in Dana, T.F. (1975), 'Development of contemporary eastern Pacific coral reefs', *Marine Biology* 33: 377–374.

26 Grant, V. (1971), *Plant Speciation*, Columbia University Press, 563pp.
27 Veron, J.E.N. (2002), 'Reticulate evolution in corals', *Proceedings of the ninth International Coral Reef Symposium*: 43–48.
28 Veron, J.E.N. (1995), *Corals in Space and Time: The Biogeography and Evolution of the Scleractinia*, Cornell University Press, New York. 321pp.
29 Grigg R.W. (1995), 'Evolution by reticulation', *Science* 269: 1893–1894.
30 Vicariance is the proposition that if a gene pool is divided by a barrier over evolutionary time, two or more species may exist if that barrier is removed. Thus species are formed by division, but not fusion, of genetic lineages. Phylogeography proposes that species originate spatially by vicariance as revealed by cladistics. On small scales of space and time this should mostly be so; however, cladistics disguises evolutionary change on larger scales.
31 Veron, J.E.N. (2008), *A Reef in Time: The Great Barrier Reef from Beginning to End*, Belknap, Harvard: 289pp.
32 Darwin, C.R. (1842), *The Structure and Distribution of Coral Reefs. Being the First Part of the Geology of the Voyage of the Beagle, Under the Command of Capt. Fitzroy, R.N. During the Years 1832 to 1836*, London: Smith Elder and Co.
33 Letter from Sir Charles Lyell to Sir John Herschel, 1837.
34 H.S. Ladd, E. Ingerson, R.C. Townend, M. Russell and H.K. Stephenson (1953), 'Drilling on Enewetak Atoll, Marshall Islands', *American Association of Petroleum Geologists Bulletin* 37: 2257–2280.
35 International Consortium for Great Barrier Reef Drilling (2001), 'New constraints on the origin of the Australian Great Barrier Reef: results from an international project of deep coring', *Geology* 29: 483–486.
36 Beaman, R.J., www.deepreef.org
37 Veron, J.E.N. (1985) 'Aspects of biogeography of hermatypic corals,' *Proceedings of the Fifth International Coral Reef Symposium, Tahiti*, 4: 83–88.
38 Veron, J.E.N. and 6 co-authors (2009), 'Delineating the Coral Triangle', *Galaxea* 11: 91–100.
39 Veron, J.E.N. (2000), *Corals of the World*, Australian Institute of Marine Science (3 volumes): 1410pp.
40 Veron, J.E.N. (2006), 'Darwin Medal presentation: corals – seeking the big picture', *Coral Reefs* 25: 3–6.
41 Aragonite, the form of calcium carbonate that modern corals use for building skeletons, is more soluble than other forms. This makes corals relatively vulnerable to ocean acidification. Palaeozoic corals built their skeletons of calcite, the least soluble form, but they went extinct at the end of the Palaeozoic anyway.

42 Veron, J.E.N. and nine co-authors (2009), 'The coral reef crisis: the critical importance of <350ppm CO_2', *Marine Pollution Bulletin* 58: 1428–1437.
43 Veron, J.E.N. (2009), 'The coral reef crisis', *Occam's Razor*, ABC radio.
44 Veron J.E.N., Stafford-Smith, M.G., DeVantier, L.M., and Turak, E. (2015), 'Overview of coral distribution patterns of zooxanthellate Scleractinia', *Frontiers in Marine Science doi*: 10.3389/fmars.2014.00081.
45 Veron J.E.N., Stafford-Smith, M.G., Turak, E., and DeVantier, L.M., (2016), www.coralsoftheworld.org.
46 Veron, J.E.N. (2015), 'The potential of type species to destabilise the taxonomy of zooxanthellate Scleractinia,' *Zootaxa* 4048: 433–435. 42.

Acknowledgements

Many who read this book might well wish that various of their friends and family had also recorded their life story and I hope some will be encouraged to do so themselves, for their loved ones at least. I wrote an account of my life, calling it 'Charlie Veron's Story', for this reason. All well and good, but a private family memoir is a very different matter from publishing one for the whole wide world to read. Being a private person, I agonised over this for years and was still prevaricating when I visited Rick Smyth, my friend since early childhood and his wife Jane. Janey put it simply: 'Charlie, when you're eighty, are you going to be happy this was published or not?' So I stopped dithering and called Iain McCalman, who had used 'Charlie Veron's Story' when writing his extraordinary book *The Reef: A Passionate History*. Iain referred me to Ben Ball, publishing director at Penguin Random House, who gave me some very sound advice about how the manuscript might be redrafted. I thank Ben for all he did. He then turned it over to Meredith Rose, who Iain, always one for a joke, described as 'a real dragon'. As expected, the dragon turned out to be as un-dragon-like as could possibly be imagined. Thank you so much, Meredith, for all you did for me, and for just being you.

As I have said in all my books over the past two decades, my partner, Mary Stafford-Smith, has always given me compelling advice. This book is no exception. I thank her again for her insights, thoughts and corrections. Without the encouragement and help of Iain, Ben, Meredith and Mary this memoir would never have come into being. I have been very privileged.

I thank Geoff Kelly for his drawings and paintings. His artwork is everywhere in my books, all given freely and always exceeding what

ACKNOWLEDGEMENTS

I hoped for. Once again, I feel very grateful to that wonderfully talented and thoughtful man. I wish there had been space for more of his work here.

Memoirs with the time-span of this one are very demanding, especially when it comes to recollecting details of events long ago and more particularly putting them into some sort of chronological context. Kirsty, my former wife, has an excellent memory for such matters and after reading my original manuscript was always there to help, encourage and correct.

Back in 2005, Gregg Borschmann, now environmental reporter for our national radio, made about eight hours of recordings of me chatting for Australia's oral history and has since interviewed me many times for various ABC radio programs. All this helped me draw the strings of my complicated life together and I am grateful to him for doing so.

Helped by Gregg no doubt, a string of writers, journalists, educators, filmmakers and television presenters have introduced me as 'a modern Darwin' or 'the godfather of corals' or the like. Not surprisingly, many of these people have wanted to know about my background and there have been others, most memorably Sir David Attenborough, who encouraged me to write this book. Of course, I found their interest flattering but more importantly, encouraging, when my many doubts about doing such a thing needed propping up.

I resisted the temptation to send the manuscript to a lot of people. After all, I hate killing trees and my story goes down many paths that they, individually, would know little about. However, I did send it to Richard Pearson, Len Zell, Lyndon DeVantier, Mike Balson, Katie Veron and Ric How. I appreciate their corrections, comments and reminiscences. My links with the Birtles family are many; those with Hillary Birtles' parents go back my entire life, so they alone were able give me an adult's glimpse of both me and my parents from my earliest years.

Most of the people mentioned in this book, and dozens who are not, have had emails or phone calls from me when I wanted to know a detail about a person, a place or an event. Although they may not have known it at the time, Terry Done, John Meagher, Hal Heatwole, Rich Pyle, Kerry McGregor, Dave Hannan, Emre Turak, Mac Horn and Susan Kennedy

ACKNOWLEDGEMENTS

all contributed over the years in different ways.

I'm grateful to the photographers who have allowed me to use their work. I especially thank Peter Donkers for the extraordinary aerial photo of the Long Reef rock platform where I played as a child, and to Phil Colman for alerting me to it, the cover photo of his own book on Australia's temperate seashores. Special thanks go to John Rotar for digging out his old photos taken during our work at the Solitary Islands during our earliest years of scuba diving. I also thank my old friends Boris Preobrazhensky, Rick Grigg and David Stoddart who died while this book was being written.

In deep retrospect, I pay homage to my parents, Don and Mary, with belated apologies for all the trouble I caused. Also to Noni and to Jinka for playing such a big part in my life before my present family took over.

Index

Aboriginal peoples 18, 139, 150, 210, 293
acidification *see* ocean acidification
Acanthastrea lordhowensis 147
Acropora 169
Africa 170, 197, 200, 213–4, 234
Agassiz, Alexander 237
AIMS
    administrative structure 107, 130, 137–8, 249, 253
    bureaucracy 106, 130, 136–8, 169, 192, 251–3, 290–1
    Cape Ferguson 136
    Cape Pallarenda 105–6, 134, 136
    chairmen 106, 130, 137–8, 148, 253
    chief scientist and resignation as 251–2
    climate change, policy about 251
    computer technology 209, 217, 230–1, 240–2, 247, 236–4, 270, 289
    conflicts with 129–30, 139, 169, 249, 251
    Corals of the World Online (website) 264
    council 106–8, 135–7, 139–40, 169, 247, 249, 251, 253
    decline of 249, 252–3
    directors 105–6, 139–41, 247, 249–52
    employment by 105–6
    executive and resignation from 249
    first scientists 105–6
    first staff 108
    history of 107–8, 136
    managers 106–8, 137
    promotions 138, 169, 192, 288
    reflections on 252–3
    relocation of 136
    resignation from 252
    scientists 137–8, 251–2
    scuba diving 141, 277–8
    tenure at 169
AIMS: The First Twenty-five Years (book) 253
algae
    kelp 2, 4–5, 8, 281
    Sargassum 173
    symbioses 87, 89–90, 254–5
    zooxanthellae 87, 89–90, 245–5
Agassiz, Alexander 237
Álvarez, Victoriano 222
amethystine python 53, 55
American Academy of Underwater Sciences 281
anemones 1–3, 6, 58, 84, 88–9, 217
anoxia 204, 257
    and ocean acidification 217, 256–8
Antarctic Circumpolar Current 146, 235
Antarctica 21, 213, 234–5, 257
Anthropocene 259
aptitude tests 36–7
aquaria 6, 8, 77, 104, 212, 275
aquarium industry 275–6
aragonite 258, 299
A Reef in Time (book) 234, 239, 253, 257–9
    publication of 253
    research in France 250
    value of 261
Arrawarra Headland 58, 61
Arrawarra marine station 58
Ashmore Reef 171
Asia-Pacific Economic Cooperation 243
Askin government, corruption 131, 147

INDEX

asteroids 234–5
asthma 24, 39, 42, 51, 282, 286
   and diving 282, 286
Atlantic Ocean
   and dispersal 214–5
   paleotemperature 214
atolls
   drilling of 237–9
   and sea level change 236
   structure of 233, 236–8
   thickness of 238–9
atolls, origins of
   Agassiz's views 237
   Darwin's theory 233, 236–9
   Lyell's views 236
Attenborough, David
   at Heron Island 292
   interviews with 292
   at Rivendell 291
   at the Royal Society, London 260–1
Australia
   and continental drift 25, 225, 234–5
   first reefs 235
   geological origins of 25, 234–6
   origin of corals 235–6
   origin of marine life 212, 235
   origin of terrestrial life 25, 235
   and paleoceanography 235
   and plate tectonics 234–6
Australian Security and Intelligence Organisation (ASIO) 116
Australian Institute of Marine Science *see* AIMS
Australian Seashores (book) 11, 13
azooxanthellate corals 89

Back, Ken 83, 92, 105, 155, 158
Back, Pat 105
Barker College 21–35, 37
Basten, Henry 148
Batavia 174
Belize 212, 215–6
Bell, Peter 253
Bennett, Isobel 11–13, 69, 82, 105, 158, 207

Berlin museum (Museum für Naturkunde) 119, 124, 181, 198
Berlin Wall 124
Bernard, Henry 122–3, 267
binomial nomenclature 231, 267, 273
biogeography 127–8, 169–70, 189, 191, 97, 205, 209, 221, 224, 240–1, 263, 289–90
Birtles, Alastair 77–8, 165
Bjelke Petersen, Joh 73, 75
bleaching, coral 255
   first record of 192
   see also mass bleaching
blue coral 184, 188, 220
blue-ringed octopus 6–7
boats and ships
   Batavia 174
   Coongoola 196
   Emden 204
   FeBrina 243
   Hero 142
   Inga Viola 201
   James Kirby 79, 81, 83, 93–6, 149–50
   jet 61
   Kallisto 109, 113–6, 118
   Lady Basten 148–9, 159
   Marco Polo 81–3, 106
   Noel Buxton 206–7
   Wanda 209–10
   wrecks 85, 142, 204
   Yongala 104
Bonnie Prince Charlie 10
boreholes in reefs 238–9
boundary currents 170–1, 190, 192, 279
branching corals 90, 254
brittle stars 2–3, 6
British Museum (Natural History) 119, 122–3, 125
bullying, school 22–4
Bunt, John 241
Burdon-Jones, Cyril 76–7, 82–3, 106–8, 154, 267
bushwalking 17–8, 33, 36, 43–4, 46, 51–2, 288
Büstt, John 72

305

INDEX

Canning Basin Devonian reefs 257
Canning Stock Route 210
canoeing 52
carbon dioxide
   and anoxia 204, 257
   and the carbon cycle 254, 258
   and carbonic acid 214, 254, 256–8, 258
   Cenozoic 235
   and climate change 214, 217, 254, 258–9, 261
   and coal mining 256
   and coral growth 213–4, 258, 261
   and global warming 213, 254
   and greenhouse effect 254
   and mass extinctions 213–4, 217, 235, 256–7, 259
   Miocene 213
   and ocean acidification 217, 256–8
   and ocean temperature 213, 254, 258
   Palaeozoic 217, 257
   and plate tectonics 213, 235–6
   and reefs 213, 256, 261
   in seawater 217, 254, 256, 258
carbon reservoirs 254, 258
Caribbean
   corals 212–7
   freshwater inundation of 216
   geological history 92, 212–5
   isolation of 213–5
Carolan, Laura 248
Cenozoic
   carbon dioxide 235
   corals 217–8, 235
   extinctions 213–4, 217, 235, 259
Chagos 203
Chappell, John 86
Chasing Coral (film) 292
chief scientist (AIMS) 251–2
chess 42–3, 114–5, 241, 287
Chesterfield Islands 148–9
Chevalier, Jean-Pierre 124
Chumbe Island Coral Park, Zanzibar 200–1
CITES 275–6
cladistics
   and evolution 229–30, 233
   and syngameons 232
   and taxonomy 230, 232–3
cladograms 229–31
Clerke Reef, Rowley Shoals 172
climate change
   AIMS policy on 251
   and carbon dioxide 214, 217, 254, 258–9, 261
   and coal 256
   and coral reefs 191, 243, 250, 256, 259–61
   denialists 250, 251, 292
   and El Niño cycles 261
   and greenhouse effect 256
   and Howard government 251
   and mass bleaching 231, 259–62, 276
   media presentations about 250–1, 259–62, 292
   predictions 261, 292–3
   study of 250–1, 253, 261, 262
climbing 44–6, 257
Clipperton Atoll, Mexico 220–3
   corals 220–2
   and Cousteau, Jacques-Yves 221–2
   and human history 222
   and shark finning 221
   and sharks 221
Clipperton: The Island Time Forgot (film) 222
Cocos (Keeling) Islands 203–5
Coll, John 155
Collins, Mrs (teacher) 9–11, 28, 207
computers, early use of 217, 240–2
Conflict Atoll 111–2
conscription 50–1, 116, 195
Conservation International 242
continental drift 25, 212–5, 225, 234–6
Convention on International Trade in Endangered Species *see* CITES
Coongoola 196–7
coral(s)
   abundance 169
   aggression 90–1

INDEX

algae interactions 173
azooxanthellate 87, 89
biogeography *see* biogeography
bleaching *see* bleaching; mass bleaching
blue 184, 188, 220
boom-and-bust cycles 217
branching 90, 254
Cenozoic *see* Cenozoic
collecting 60, 79, 129, 143, 147, 184–5, 198, 222, 266, 275–6
collections 122, 124–5, 134, 141, 170, 198, 202, 220, 245, 263, 266, 268, 270, 290
colonies 80, 86, 89, 91, 146–7, 173, 189, 295
competition 87, 90–1
conservation 72–5, 95, 107, 188, 198, 233, 242–3, 260, 265, 268, 276
dispersal *see* dispersal
diversity *see* diversity
contour maps of *see* mapping coral
and environmental variation 80, 122, 129, 216–7, 267–8
drop-out sequence 178, 189–91
evolution *see* Darwinian evolution; reticulate evolution
extinctions *see* extinctions
family tree *see* family tree, Scleractinia
feeding 89–91
food sources 90
fossils 18–21, 128, 187, 206, 217–20 233–4
geographic variation in *see* geographic variation in corals
geological longevity of species 184, 188, 220
growth forms 80, 122, 135–6, 175–6, 266
growth rates 90, 258
hermatypic 89
high latitude *see* high-latitude corals
historical collections 119, 122–4
larvae 191, 205, 214, 221, 225, 227, 258
latitudinal attenuation 178, 189–91
longevity of species 184, 188, 220

mass bleaching *see* mass bleaching, coral
massive 90
mass spawning 91, 221
Mesozoic 213, 217–8, 234–5
monographs *see* monographs, coral
names 69, 80, 118–9 122, 125, 129, 147, 224, 245, 265–7, 269–74, 290
nomenclature 118–9, 224–7, 231, 265, 267, 269–74
Palaeozoic 217, 257
polyps 89, 91
reproduction 91, 221
Scleractinia 217, 220, 275
skeletons 86, 88–9, 217, 258
spawning 91 221
speciation *see* Darwinian evolution; reticulate evolution
studies, history of 266–9
symbioses 87–90, 254–5
taxonomy *see* coral taxonomy
temperature limits, lower 190–1, 235
temperature limits, upper 175, 255
territoriality 90–1
type specimens 118, 123–5, 266, 270, 272–5
zooxanthellate 89–90, 214, 221, 255
Coral Geographic computer compilations 241, 249
Coral ID (CD-ROM) 248
Corallium 275
coral reef symposia 81–2, 100–2, 117, 218, 240, 248, 260–1, 266, 281
Coral Sea 148, 196
coral taxonomy
and Darwinian evolution *see* Darwinian evolution
environmental influences 80, 122, 129, 216–7, 267–8
field work 269
history of 266–9
and the ICZN 271–4
importance of 269–70
monographs *see* monographs, coral
and museums 135, 267–8, 270, 272

INDEX

  see also museums
  problems with 176, 188, 272–4
  reservations about 176, 188
  and reticulate evolution 223–33
  and taxonomists 122, 263, 268–69
  types of 268–69
coral reefs
  and anoxia 204, 257
  beauty of 47, 83–5, 268–9
  and carbon dioxide *see* carbon dioxide
  dangers on 94, 142–5, 280–1, 286
  karst erosion 257
  mass bleaching *see* mass bleaching
  microbes 257
  noise 280, 284
  and ocean acidification 217, 256–8
  photography of *see* photography
  symbioses 87–90, 254–5
  temperatures *see* temperature
  working on 277–8
  see also reefs
Coral Triangle 200, 228, 240–4
Coral Triangle Initiative
  and *Corals of the World* website 244, 264
  early history 243
  symposium on 264
Corals of Australia (book) 191–2
Corals of Japan (book) 188, 246
Corals of Japan (monograph) 188
Corals in Space and Time (book) 217, 221, 228, 241
Corals of the World (book) 246–9, 263
Corals of the World (website) 248, 263–6
Corals of the World Online (AIMS website) 264
Cornell University 127
Cousteau, Jacque-Yves 221–3
Cowen, Zelman 140
crabs
  of childhood 3–4, 6–7
  coconut 111
Cretaceous 213, 234–5
Croll, Ian 104
crown-of-thorns starfish 66, 176, 183

CSIRO 57, 107, 138, 247
cunjevoi 8, 12
currents
  Antarctic Circumpolar 146, 235
  boundary 170–1, 190, 192, 279
  deltaic reefs 96–8, 143
  East Australian 146
  and dispersal 289, 291
  and evolution 225
  Humboldt 215
  Kuroshio 187, 190, 192, 279
  Leeuwin 170–1
  North Equatorial Counter 221
  palaeo 213–5, 221, 225, 235–6
  of the Pompey Complex 142–3
  surface 91, 225–7
  of the Tateyama Peninsula 187
  tidal 142–3, 172, 175
cuttlefish 174
cyclones 76, 85, 92, 96, 202, 295

dadirri 17–8, 211
Dakin, William 11
Dana, James 125
Dana, T.F. 221
Dangar Falls 45
Darwin, Charles
  and atoll formation 92, 123, 233, 36–9
  and evolution *see* Darwinian evolution
  and reef thickness 237–8
Darwinian evolution
  and cladistics 229–30, 233
  and natural selection 224–31
  and religion 28, 238
  and reticulate evolution 223–33
  and species concepts 224–33
Darwin Medal 248
Davies, Peter 238
Day, Max 107, 137
deltaic reefs 96–8, 172
deserts 199–210
DeVantier, Lyndon 263
  and *Corals of the World* website 263
Devonian reefs 257

INDEX

Discovery Bay Marine Lab 216
dispersal
    by surface currents 91, 225–7
    evolutionary role 225–7
    long-distance 203–5, 214–5, 221, 227, 239, 191
    on pumice 221
diversity, coral 84, 45, 171, 189–90, 197, 202–4, 213–5, 222, 241–2
diving
    computers 60, 94, 279
    rebreathers 282–3
    see also scuba diving
drop-out sequence, coral 178, 189–91
DNA
    and species longevity 220
    and taxonomy 229–30, 232, 269–70
    and type specimens 273–4
documentaries 159, 244, 251, 261–2, 291–2
dogs 16–8, 20, 29, 33–4, 36, 43, 66, 106, 133, 151, 155, 166, 228, 294
Done, Terry 60, 62, 107–8, 158, 196
dragonflies 17, 65, 67–9
drilling reefs *see* boreholes in reefs
drop-out sequence
    Australia 189
    Japan 178
Duncan, Davie 83, 94–7, 149–51

East Australian Current 146
Egypt
    Ras Mohammed National Park 197–8
    Sinai Peninsula 197–8, 199
Ehrenberg, Christian 124, 198
Eilat, Israel 206
El Niño cycles and mass bleaching 255, 261
Ellison Reef, Great Barrier Reef 72
Emden, wreck of 204
Endean, Bob 73
Enewetak Atoll 128–9
entomology 64–5
Era Beds, Papua New Guinea 219–220
eucalypts 108, 226–7, 232
eureka moments 223, 236

evolution
    and currents 225
    Darwinian *see* Darwinian evolution
    and natural selection 30, 224–31
    reticulate *see* reticulate evolution
extinctions
    of coral 214, 229
    by fusion of lineages 229, 233
    mass 217, 235, 256, 259
    and punctuated equilibria 233
    by termination of lineages 229

family tree, Scleractinia 217–9
far eastern Pacific corals 221–2
Favia and type species 274
FeBrina 243–4
Feely, Dick 258–9
field work, challenges of 267, 269–70
filter feeders 84, 90
fish ponds 14–5, 20, 52–3
Fletcher, Harold 21
fossils 18–9, 21, 128, 187, 206, 217–20, 233
France
    children's schooling 249–50
    studies in 250
freedom, importance of 18, 39, 42–3, 69, 207, 288, 291
fringing reefs 171, 179, 240
Fungia 189
funnel-web spiders 9, 19–20, 53
fuzzy boundaries *see* reticulate evolution

geographic variation in corals 216, 223, 231, 233, 263, 270
Gilmartin, Malvern (Red) 105–6, 108–9, 146
global warming *see* climate change
Glossopteris 21, 234
Glover's Reef 212
Gomez, Ed 192, 194
Gondwana
    breakup of 21, 213, 234
    daylight regimes 234
Goniopora norfolkensis 245

INDEX

Goreau, Tom 216
Grant, Verne 224
Great Barrier Reef
  age of 238–9
  bathymetry 239
  battle to protect 66, 72–5
  deltaic reefs 96–8
  diversity of 84, 145, 171, 189
  diving on 84–6, 96–8, 134, 150, 280–1
  expeditions to 81, 83, 92–101
  first dives 47–8, 57–8
  geological history 238–9
  geomorphology 238–9
  mining on 72
  navigation on 92, 97, 142
  oil drilling on 39, 66, 72–4
  origins of 235–36
  political history 72–5
  primordial 235–6
  and the Queensland Trough 92–3
  research stations 47–8, 56, 93, 149, 154–5, 292
  ribbon reefs 92–5
  royal commission into 74
  size of 83–4
  thickness of 238–9
  World Heritage listing 146, 242
Great Barrier Reef Marine Park
  history of 72–5
  proclamation of 75
  size of 83–4
  threats to 66, 72–5, 176
Great Barrier Reef Marine Park Authority 238, 242, 260
Greater Australia 235
greenhouse effect 254
groupers 85, 104, 148–9
Gulf of Mexico 216, 235
  freshwater inundation 216

Hannan, David 265
Hawaii 125, 128, 187, 266
Heatwole, Harold 53–55
Hero (boat) 142

Heron Island
  drilling 238–9
  marine station 47, 56, 149
high-latitude corals
  Houtman Abrolhos Islands 173
  Japan 186–7
  Lord Howe Island 146–7
  Solitary Islands 58–9
  Tateyama Peninsula 186–7
Hinchinbrook Island 45–7
Hodgson, Gregor 193–4
Hoegh-Guldberg, Ove 255
Homo sapiens phylogeny 228
Hong Kong coral workshop 165
Honshu 186–7
Houtman Abrolhos Islands 171, 173–5
How, Ric 42–4, 46, 48, 52, 62, 170
human evolution 228
humanity, future of 293
Humboldt Current 215
hybridisation 224, 227–8, 232–3

ice ages 86–7, 92, 139, 197, 215, 221, 237, 257
ice shelves 86 215, 237
Indian Ocean
  centre of diversity 197, 199–200
  coral dispersal 197, 200
Indonesia 170, 172, 178, 194, 242
Indonesia-Philippines archipelago 200, 242
Inga Viola (boat) 201–2
Inge (book) 156–8
Inge Moessler *see* Moessler, Inge
Intergovernmental Panel on Climate Change 261
  disagreement with 261
International Commission of Zoological Nomenclature (ICZN) 271
International Code of Zoological Nomenclature (book) 271
  historical perspective 271–2
  Latin, use of 273
  problems with 272–4
intertidal corals 175

INDEX

Ishigaki Island
  airport scandal 185–6
  banquets 180–3
  blue coral 184, 188
  corals of 182–3, 186
  disputes 184–6
  diving 182, 184–5
  mass bleaching 183
  mayor of 179–82, 185
  media appearances 182, 184
  press conferences 179, 186
  riots 183, 185–6
  Shiraho Lagoon 183
island(s)
  Chesterfield 148–9
  Chumbe 200
  Clipperton *see* Clipperton Atoll
  Cocos (Keeling) 203–5
  Coral Sea 148, 196
  Enewetak Atoll 128–9
  Heron *see* Heron Island
  Hinchinbrook 45–7
  Houtman Abrolhos *see* Houtman Abrolhos Islands
  Ishigaki *see* Ishigaki Island
  Lizard 81, 93, 149, 208
  Lord Howe 145–7
  Marshall 128–9, 237–8
  Mer 99, 206
  Murray 99, 206
  Norfolk 245–6
  Orpheus 134, 149–55, 160
  Palm 47, 134, 149–55
  Raine 95–6, 207
  Ryukyu 178, 186, 190
    see also Ishigaki Island
  Solitary *see* Solitary Islands
Israel 205–6
  Eilat 206
Isthmus of Panama 215

Jackson, John 221, 246, 248
James Cook University 77, 106–8, 116, 135
  post-doc 66, 69–70

James Kirby (boat) 79, 81, 83, 93–6, 149–50
Japan
  coral, latitudinal limits 186–7
  coral, low temperature tolerances 190–1
  currents 187, 190, 279
  diving 179, 182, 184–5, 190
  Honshu 186–7
  Kyushu 186
  ocean temperatures 190–1
  Okinawa, governor of 186
  Tateyama Peninsula 186–7
  see also Ishigaki Island
Japanese corals
  field work 182–4, 186–7
  Honshu 186–7
  Kyushu 186
  mass bleaching of 183
  publications about 188
  Ryukyu Islands 178, 186
  Tateyama Peninsula 186–7
  taxonomic problems 188
  and temperature 190–1
Jinka (dog) 16–8, 29, 33–4, 36, 66
Jordan 206
journalists 261–2

Kallisto (ship) 109, 113–6, 118
karst erosion 257
Kelley, Russell 219
Kelly, Geoff 67, 240
kelp 4–5, 8, 281
Kimberley coast 175
Kirby, Justice 147
kookaburras 131, 166, 226–7, 232
Kühlmann, Dieter 124, 181–2, 184, 186
Ku-ring-gai Chase 17, 288
Kuroshio 187, 190, 192, 279

Lady Basten (boat) 148–9, 159
larvae, coral 191, 258
  long-distance dispersal *see* dispersal, long-distance
Latin, use of 79, 273
latitudinal limits *see* high-latitude corals

INDEX

and climate change 191
and temperature 190–1
Leeuwin Current 170–1
Le Gay Brereton, John 61
light, and coral growth 87, 89–90, 191, 237, 254–5
limestone, reef 72, 88, 92, 94, 96–7, 139, 214, 234, 257, 295
Lindfield East Public School 9–11
Lizard Island Marine Station 81, 93, 149, 208
longevity, species 220
Long Reef rock platform, Sydney 1–6, 8, 11, 288
Lord Howe Island 146–7
Lousisiade Archipelago 112
Lovell, Ed 171, 192
Lyell, Charles 236
lyrebirds 44

Mabo, Eddie 99–100
Mackenzie, Kirsty *see* Veron, Kirsty
Maclay, Nikoli 115
Madagascar 202–3
malaria 284–5
Maldives 203
manta rays 47, 85
mapping coral
    diversity 197, 199, 241–2
    genera 197, 240
    species 199, 241–2
    see also Corals of the World (website)
Marco Polo (ship) 81–3, 106
marine parks
    Coral Sea 148–9
    Great Barrier Reef *see* Great Barrier Reef Marine Park
    Ningaloo Reef 170–1
    Solitary Islands 58, 62
marine stations
    Arrawarra 58
    Discovery Bay 216
    Enewetak 128
    Heron Island 47, 56, 149

Lizard Island 81, 93, 149, 208
Orpheus Island 149, 155
Marshall Islands 128–9
mass bleaching 183
    aftermath of 255–6
    cause of 255
    and coal 256
    coral 183, 255–6
    ecological consequences of 255–6
    and El Niño cycles 255–61
    geological record 255–6
    and light 89, 255
    and ocean acidification 217, 256–8
    and temperature 255
mass extinctions 217, 235, 256, 259
massive corals 90
mass spawning 91, 221
Mather, Pat 81, 135, 271
Matthai, George 267
medical aspects of diving 282–6
meditation 17–8, 211
Mediterranean Sea 87, 197
    geological origin of 197, 213–4
Melbourne Ward, museum 18, 115
Melville, George 106, 137
Mer Island 99–100, 207
Mesoamerica 215, 221
Mesozoic corals 213, 217–8, 234–5
Mesozoic Era 217
Milankovitch cycles 87
Milne Bay, Papua New Guinea 242
mining, Great Barrier Reef 39, 70, 72–3
Miocene and Australia 214, 235–6
Miocene corals, origins of 197, 214–5, 235
Miocene reefs
    origins of marine life 197, 213–5
    and tectonic movements 213–4
Moessler, Inge 151–8
    daughter, Ingrid 156–7
molecular taxonomy 269–70
monographs, coral 79, 81, 103, 106, 118, 134–6, 169, 188, 209, 267, 274
monsoons 92, 94, 141, 152
Montipora 125, 169

INDEX

Moorea, influence on Darwin 236
Murray Island 99–100, 207
Muscatine, Len 89
museum taxonomists 135, 267–8, 270, 272
Museum of Tropical Queensland 81
museums 266, 270–1, 289
    Berlin 119, 124, 198
    British (Natural History) 119, 122–3
    Comparative Zoology, Harvard 237
    Melbourne Ward's 18, 115
    Paris 119, 124
    Queensland 81
    Smithsonian Institution 119, 125–6, 185, 274
    of Tropical Queensland 169
    Western Australian 170–1, 203
    Yale, Peabody 125
Muzik, Katy 185

National Geographic (magazine) 20, 25, 36, 195
Nature, importance of 9, 53, 153, 289, 293
    in children 9, 14, 17, 40, 133
    in education 40, 289
    in scientists 158, 226, 289
Nature Conservancy, The 243, 258
navigation
    GPS 172
    Great Barrier Reef 92, 97, 142
    by sea gypsies 172
Neanderthal man 228
Nemenzo, Francisco 193
Neo-Darwinian evolution 224, 228
    see also Darwinian evolution
night diving 84–5, 91, 95, 280–1
Ningaloo Reef 170–1
Nishihira, Moritaka 186, 188
Noel Buxton (boat) 206–7
noise of coral reefs 84
nomenclature 267–8
    coral 268–9
    faults of 272–4
    and fuzzy boundaries 231
    history of 271–2

    and the ICZN 271
    importance of 269–70
    misuse of 272–4
    and synonymy 231, 263, 268
Norfolk Island 245
North Equatorial Counter Current 221

O'Farrell, Professor 57, 64, 68, 70, 76
ocean acidification 217, 256–8
Ocean Ark Alliance 265
ocean currents *see* currents
Oceania House 204
Occam's Razor (radio program) 262
octopus, blue-ringed 6–7
Octopussy (film) 7
oil drilling, Great Barrier Reef 39, 66, 72–4
Oken, Lorenz 273
origins of
    atolls *see* atolls, origins of
    Australian corals 235–6
    Caribbean corals 212–5
    Great Barrier Reef 235–6
    Indian Ocean corals 197, 199–200, 212–5
    Pacific corals 189, 221, 215, 227
        see also Coral Triangle
    Red Sea corals 197, 200, 213–4
    Western Australian reefs 170
Orpheus Island Research Station 149, 155

palaeontology 21, 218–9
paleoceanography *see* currents, palaeo
Palaeozoic Era 217, 257
Palm Islands 47, 134, 149–50
Papua New Guinea 86
    Conflict Atoll 111–2
    Era Beds 219–20
    Louisiade Archipelago 112
    Milne Bay 242
    Port Moresby 219
    Sepik River 115
Paris, museum 119–124
Pearce-Kelly, Paul 259–60

INDEX

and the Royal Society, London 260–1
Persian Gulf 175
Philippines 192–5
photography
  aerial 93, 96–79, 143–4, 183
  underwater 47, 143, 147–8, 174, 182, 184, 192, 198, 204, 216, 246–8, 292
phylogeny, primate 229, 230, 270
phylogeography 231, 299
Pichon, Michel 135
Pioneer Bay, Palm Islands 150, 152, 154–5
plate tectonics 25, 212–5, 225, 234–6
Pleistocene 139, 215, 294–5
  corals 219–221
  ice age 86–7, 92
  reefs 86–7, 92
  sea levels 86–7, 92, 139, 197, 237, 257
Plesiastrea russelli 129
Pocillopora damicornis 272
polychaete worms 1–2, 4–6, 8, 10–12, 59
Pompey Complex, Great Barrier Reef 142–3
Porites 169, 189, 219, 222
Port Moresby 219
possums, study of 44
post-doc, James Cook University 66, 69–70, 76, 83, 267, 288
Preobrazhensky, Boris 110–1, 113, 115–8
punctuated equilibria 233

Queensland grouper 148–9
Queensland Museum 81
Queensland Trough 92–3, 233–5

Raine Island 95–6, 207, 292
Ras Mohammad National Park, Egypt 197–200
rebreathers 282–3
Red Sea, geological history 197, 214
Red Sea corals 124, 197, 200, 213–4
  diversity 199
  origins of 197–200, 212, 214
  studies of 197
Reef in Time, A (book) 234, 239, 253, 257–9

reef(s)
  aquaria 104, 147, 275
  Ashmore 171
  builders of 80, 86, 88–90
  Caribbean 212–5
  Cenozoic 213–4, 217, 235, 259
  Chesterfield 148–9
  Chumbe Island 200
  Clerke 172
  Cocos (Keeling) Island 203–5
  deltaic 96–8
  Devonian 257
  diversity 145, 171, 189–90, 197, 202–4, 213, 215, 222, 241–2
  drilling on 39, 66, 70, 72–4, 238–9
  Ellison 72
  Enewetak 128–9
  first Australian 238–9
  flats 87, 93–4, 144, 182–3, 185, 266
  fossil record of 217–8, 214, 234, 257
  fringing 171, 179, 240
  geological reconstructions 92, 238–9
    see also reefs, drilling
  geology 86–8
    see also reefs, drilling
  Glover's 212
  Great Barrier see Great Barrier Reef
  greater Australia 235
  Heron Island 47, 56, 149, 238–9
  Houtman Abrolhos Islands 173–5
  Indonesia-Philippines archipelago 200, 242
    see also Coral Triangle
  Ishigaki Island see Ishigaki Island
  Mesozoic 213, 217–8, 234–5
  Miocene 197, 213–5
  morphology 92, 123, 233, 236–9
  Ningaloo 170–1
  Papua New Guinea 86, 111–2, 219–20
  Pompey Complex 142–3
  and the Queensland Trough 92–3, 233–5
  ribbon 92–5
  Rowley Shoals 171, 173
  Scott 171

INDEX

Solomon Islands 243–4
  thickness of 238–9
  Tijou 92–5
Reichelt, Russell 247
reproduction, coral 91, 221
reproductive isolation 224–9, 232–3
research stations
  Discovery Bay 216
  Heron Island 47, 56, 149
  Houtman Abrolhos Islands 174
  Lizard Island 81, 93, 149, 208
  Marshall Islands 128–9
  Orpheus Island 149, 155
research vessels *see* boats and ships
reticulate evolution
  and cladistics 229–30, 233
  of corals 227
  and Darwinian evolution 229–30, 233
  and environments 226–7
  fuzzy boundaries of 231
  and humans 228
  and hybridisation 224–5, 228
  name, use of 224–5
  and reproductive isolation 224–9, 232–3
  and sea surface currents 225–6
  and syngameons 225, 232–3
ribbon reefs
  age of 92–3, 235
  bathymetry 92, 233
  drilling on 238
  formation of 93, 235
  and the Queensland Trough 92
Riedmiller, Sibylle 200
Riegl, Bernard 199
Rivendell 129, 134, 157–60, 163–4, 166–7, 206, 208, 211–2, 240, 245–6, 250, 260, 283, 287, 291, 294–6
Rotar, John 59, 62, 177
Rowley Shoals 171, 173
Royal Society, London
  and Attenborough, David 260
  and the Stoddart expedition 100–2
  and author's presentation 100–2

Russia and Russians
  Amur 117
  border guards 117, 124
  and *Kallisto* 109, 113–6, 118
  KGB agents 113
  Khabarovsk 117, 124
  Preobrazhensky, Boris 110–11, 113, 115–8
  Sorokin, Yuri 111, 115
Ryukyu Islands 178, 186, 190
  see also Ishigaki Island

St Alban's Church, Sydney 13, 29, 120
Sargassum 173
Science Show, The 139
Scleractinia
  Cenozoic 217–8, 235
  family tree of *see* family tree, Scleractinia
  first 187, 218
  fossils 217–20
  Mesozoic 213, 217–8, 234–5
Scleractinia of Eastern Australia (monographs) 134, 168, 189, 192
Scott Reef 171
scuba diving
  AIMS 141, 278–9
  alone 182, 278–82
  and asthma 286
  best dives 280–1
  best places 86, 280–1
  buddies 278–9
  and computers 60, 94
  and conservation 267–8
  and coral collecting 143, 184–5, 286
  and coral taxonomy 122, 129, 135, 267–8
  currents 82, 85, 96–8, 142, 170, 279
  dangers of 85, 277, 286
  decompression 280
  deltaic reefs 97–8
  early scientific use of 216, 221, 292
  early equipment 60, 104, 112
  first dive 57
  injuries 142
  Ishigaki Island 181–5
  Japan 181–5, 186, 278–9

INDEX

joy of 286–9
medical hazards 282–4, 286
at night 84–5, 91, 95, 280–1
officer, AIMS 141
officer, CSIRO 57
officer, James Cook University 80
and post-doc, James Cook University 66–7, 79–81, 94, 267
rebreathers 282–4
regulations 79–80, 277–9, 281–2
research, used in 58–62, 66–7, 79–80, 94, 97–8, 112–3, 129, 134, 144–5, 152, 192–3, 201–2, 216, 266–7, 277–80, 286–7
ribbon reefs 92–3
Solitary Islands 58–62
tables 279–80
and taxonomic studies 80, 122, 129, 135, 267–9
sea gypsies 172–3
sea levels
and atoll formation 237
changes in 86–7, 92, 96, 98, 139, 254, 257
and climate change 87, 92, 96, 97
effect on reefs 87, 92, 98, 139, 237, 239, 254, 257, 295
future predictions 257
and the Great Barrier Reef 92, 96, 98, 139, 142, 239, 257, 295
maximum rate of rise 257
sea lions 174–5
sea snakes 54–5, 143–4
sea squirts 4, 8, 12
seaweed 173, 180, 189, 191, 254–5
Sepik River 115
Seychelles 203
shark
attacks 144–5
bull 144
conservation 187, 286
finning 187, 221
great white 170
grey reef 145

hammerhead 144
tiger 95, 144
whale 85, 170–1
ships *see* boats and ships
shipwrecks 85, 99, 142, 204
Shiraho Lagoon 185–6, 188
Shirai 178–84, 186, 188
Sinai Peninsula, Egypt 197–200
  corals 198–200
  monastery 199
Smithsonian Institution 119, 125–6, 185, 274
Smyth, Rick 15–6, 46
snake(s)
  amethystine python 53, 55
  handling of 53–4
  house, University of New England 53, 55–6
  land 19, 53–7, 133, 159
  sea 54–5, 143–44
Solitary Islands 58–62
Solomon Islands 243–4
Sorokin, Yuri 111, 115
species
  concepts 122–3, 124–5, 224, 233, 266–7, 289
  as continua 224–233
  corals 80, 122–5, 227–8
  in *Corals of the World* (website) 263–6
  distinctions underwater 80, 176, 188, 267, 270
  spawning 91, 221
  as units 224–33
  longevity of 184, 188, 220
  origin of 225–30
  speciation 225–30
spiders, funnel-web 9, 19–20, 53
Sri Lanka 203
Stafford Smith, Clive 260
Stafford Smith, Mark 210
Stafford-Smith, Mary
  childhood home 208–9
  and children 210–1, 245–6, 249, 287
  computer skills 209, 242, 247

INDEX

and *Corals of the World* (book) 246–9, 263
and *Corals of the World* (website) 263
first meeting with author 208
life with 208–12, 216, 245–66, 282–7
post-doc, James Cook University 209
and *Wanda* 209
starfish 3, 6, 66, 84, 176, 183
Steedman, Ray 253
Stoddart, David
  expedition 83, 92, 100
  friendship with author 103
  Marco Polo, on the 81
  Royal Society, at the 100–2
Stonehenge station 52, 56, 70, 119
surface currents *see* currents
symbiosis, algal
syngameons 225, 232–5
synonymy of names 231, 268
  historical 263, 268
  in taxonomic disputes 263, 268
  uncertainties of 263

Tanna, Vanuatu 196
Tateyama Peninsula, Japan 186–7
taxonomists 268–71
  Bernard, Henry 122–3, 267
  Dana, James 125
  DeVantier, Lyndon 263
  Ehrenberg, Christian 124, 198
  field 267, 269–70
  history of 266–271
  interpreting past 122–3, 125
  molecular 269–70
  museum 266–8
  Pichon, Michelle 135
  Turak, Emre 263
  types of 268–71
  Verrill, A.E. 125, 274
  Wallace, Carden 169, 171, 192
  Wijsman-Best, Maya 135–6
taxonomic publications
  authors of 231, 270
  and cladistics 229–30, 233

historical monographs 79, 103, 118, 270
PhD theses 270
and reticulate evolution 265–6
and wildcards 270
taxonomy
  cladistics 230, 232–3
  coral *see* coral taxonomy
  and fuzzy boundaries 231
  and ICZN 271–4
  importance of 271
  science of 271
technology, use of 218, 263, 269–70, 289, 292
temperature
  and carbon dioxide 213, 254, 258
  and coral growth 190–1
  and diversity 190, 203
  extremes 175, 186–7, 190–1, 235
  gradients 178, 189–91, 203
  latitudinal attenuation 190–1
  and mass bleaching 255
  and mass spawning 91
  palaeo 215, 235
  ranges 190, 255
  and reef formation 190–1
  studies of 190–1
  tolerance 175, 190–1, 255
Tethys Sea 212–4
  and Austrian Alps 218
  birthplace of corals 197, 200, 212–3
  geological history 197, 212–3
  obliteration of 213–4
Thailand 192, 194, 203
tidal currents 96, 142–3, 172, 75
tidal flushing 204
tidal ranges 142, 175
Tijou Reef 92–5
Torres Strait 74, 83, 92, 189
Townsville 71, 74, 76–7, 104, 106, 108
Turak, Emre 263
  and *Corals of the World* website 263
Turbinaria 273
turtles, Raine Island 95

INDEX

type specimens 123–5, 266, 270, 272–4
  collecting of 122–3, 266–7
  historical treatment 122–4
  problems with 176, 188, 272–4
  uncertainties about 122–4

UNESCO 171
United Nations Framework Convention on Climate Change 243
universities
  Cornell 127
  James Cook 66, 69–70, 76, 83, 267, 288
  New England 38–43, 48–50, 56, 176–7
  Queensland 73

Vanuatu 196
Vaughan, Thomas 274
Veron, Charlie
  Aboriginal world, affinity with 18, 139–40, 210, 293–5
  asthma 24, 39, 42, 51, 282, 286
  and Attenborough, David 260–1, 292
  Bible, views on 13–4, 29–31
  books, love of 20
  and bullying, school 22–4
  and bureaucracy 106, 130, 136–8, 169, 192, 251–3, 290–1
  and bushwalking 17–8, 33, 36, 43–4, 46, 51–2, 288
  and canoeing 52
  career and promotions *see* AIMS; James Cook University; University of New England
  and chess 42–3, 114–5, 241, 287
  chief scientist and resignation as 251–2
  childhood 1–38
  children *see* Veron, Eviie, Katrina, Martin, Noni, Ruari
  churches and 13, 29, 120
  and climate change *see* climate change
  and climbing 44–6, 257
  and conscription 50–1, 116, 195
  Darwin Medal 248
  Darwin, thoughts about 28–9, 92, 230–1, 233, 236–8, 240
    see also reticulate evolution
  and daydreaming 14, 20, 24, 26
  degrees 49, 53, 56, 70, 176
  diving *see* scuba diving
  divorce 207–8
  and dogs *see* dogs
  and dragonflies 17, 65, 67–9
  and entomology 64–5
  eureka moments 223
  father, relationship with 14, 35, 121, 167
  freedom, views on 18, 39, 42, 69, 207, 288, 291–3
  history, love of 122
  influences on 292
  marriage 56, 206
  mechanics, love of 14, 31, 34–5, 201
  media, use of 291–2
  medical issues 282–6
  meditation 17–8, 211
  monographs *see* monographs, coral
  mother, relationship with 13–4, 26, 29, 40, 50, 79, 119, 121
  museum, childhood 18–9
  music, love of 40–3, 133, 153, 158
  Nature, affinity with 13–4, 17, 29, 33, 40, 43, 53, 133, 152–3, 158, 226, 286, 288–9, 293
  nickname 9–11, 24, 28, 42, 56, 111
  parents, relationship with *see* father, relationship with; mother, relationship with
  partner *see* Stafford-Smith, Mary
  publications *see A Reef in Time* (book); *Coral ID* (CD-ROM); *Corals of Australia* (book); *Corals of Japan* (book); *Corals of Japan* (monograph); *Corals in Space and Time* (book); *Corals of the World* (book); *Corals of the World* (website); *Inge*
  and polychaete worms 1–2, 4, 12
  post-doc (James Cook University) 66, 69–70, 76, 83, 267, 288

INDEX

and religion 13–4, 28–31, 152
resignation from AIMS 252
and Rivendell *see* Rivendell
and the Royal Society *see* Royal Society, London
school life 21–31, 35–38
sex education 16
shyness 16, 31–2, 39, 49–50, 54
solitude, love of 17–8, 33–4, 36, 43, 148, 287
speech impediment 24–8, 30
and spiders 9, 19–20, 53
and Stoddart, David *see* Stoddart, David
student life 39–43, 44–5, 48–9
taxonomy, views on *see* taxonomists; taxonomy
teachers 9–11, 22–31, 207
university
life 38–43, 48–50, 56, 176–7
scholarship 35–38
Vietnam War, avoidance of 50–51
vocational guidance 35
and *Wanda* 209
wife *see* Veron, Kirsty
wilderness, love of 17–8, 33–4, 36, 43, 148, 287
see also AIMS; scuba diving
Veron, Donald (father) 14–5, 18, 27–8, 65, 120
childhood influence 14, 18, 34–5, 121, 167
justice crusades 130–131, 147
military history 27–8
relationship with author 14, 18, 34–5, 121, 167
Veron, Eviie (daughter)
birth of 211
life with 245, 249, 287
Veron, Fiona (daughter) *see* Veron, Noni
Veron, Jan (sister) 5, 13, 20, 66, 120, 167–8, 211
death of 211
life with 5, 13, 20, 66, 120, 167–8, 211

Veron, John (author) *see* Veron, Charlie
Veron, Katrina (daughter) 161–4, 167, 205, 250, 283, 287
birth of 161
early medical history 161–2, 164, 250
life with 164, 167, 205, 245, 287
Veron, Kirsty (wife)
life with 49–52, 56, 65–8, 70–1, 76–9, 102, 105, 108, 119–24, 132–3, 146, 160–5, 205–7
marriage 56
miscarriages 78–9, 159
separation and divorce 205–8
Veron, Martin (son)
birth of 246
life with 246, 249, 287
Veron, Mary (mother)
importance of 13–4, 26, 29, 40, 50, 79, 119, 121
relationship with author 1–6, 11–17, 19, 21, 26–7, 29–30, 32–4, 38, 50, 67, 78–9
death of 120
and Kirsty 67, 78, 119–20
and Noni 67, 71, 78, 119–21
and religion 13
Veron, Noni (daughter)
birth of 65
death of 165–6
early childhood 66–8
horseriding 160
music 133, 159, 163
Nature, love of 132–3, 160
relationship with author 67–8, 78–9, 119–21, 132–4, 159
and Rivendell 109, 132–4, 160
schooling 79, 163–4
scuba diving 159–60
Veron, Ruari (son) 67
death of 68
Verrill, A.E. 125, 274
Verrilliana 125
vicariance 231
Vietnam 194–5

INDEX

Vietnam War 50–1, 116
Wallace, Carden 169, 171, 192
Wanda 209
Ward, Melbourne 18
Watkins, Cocky 142
website, *Corals of the World* see *Corals of the World* (website)
Wells, John 127, 129, 135, 189, 197, 216, 218, 240, 266, 274
whales 15, 85, 94–5
whale sharks 85, 149, 170–1
Western Australia
    corals 169–70, 175–6
    field trips 171, 173, 175
    islands 173–5
    reefs 170–2
    museum 170–1
Wijsman-Best, Maya 135–6
wilderness 148, 210–1, 287
Williams, Gary 248
Williams, Robyn 139, 262
Wilson, Barry 170
Windeyer, Justice 38
World Heritage
    Great Barrier Reef 242
    Ningaloo Reef 171
Wright, Judith 39
Wright, P.A. 39

Yongala (boat) 104
Yonge, Maurice 101–3

Zanzibar 200
    Chumbe Island Coral Park 200
Zell, Len 62, 108
Zoological Society of London 259
Zoopilus 196
zooxanthellae 89, 213 221, 255
zooxanthellate corals 89–90, 214, 255